Fundamentals of
MODERN
MATHEMATICS
A Practical Review

David B. MacNeil

Dover Publications, Inc.
Mineola, New York

CUYAHOGA COMMUNITY COLLEGE
EASTERN CAMPUS LIBRARY

Bibliographical Note

This Dover edition, first published in 2013, is an unabridged republication of the work originally published by D. Van Nostrand Company, Inc., Princeton, N.J., in 1963 under the title *Modern Mathematics for the Practical Man.*

Library of Congress Cataloging-in-Publication Data

MacNeil, David B.
    Fundamentals of modern mathematics : a practical review / David B. MacNeil. — Dover edition.
      p. cm.
    Reprint of: Princeton, N.J. : D. Van Nostrand Company, Inc., 1963. — (Mathematics for self study).
    Summary: "Students and general readers wishing to know a little more about the practical side of mathematics will find this volume a highly informative resource. Worked examples and diagrams illustrate important concepts in accessible explanations of set theory, numbers and groups, matrices and determinants, probability and statistics, game theory, and many other topics. 1963 edition" — Provided by publisher.
    Includes index.
    ISBN-13: 978-0-486-49745-7 (pbk.)
    ISBN-10: 0-486-49745-3 (pbk.)
    1. Mathematical analysis. 2. Algebra. I. Title.

QA37.M19 2013
512—dc23

2012047317

Manufactured in the United States by Courier Corporation
49745301   2013
www.doverpublications.com

PREFACE

The purpose of this book is to explain and demonstrate a broad selection of the more important developments in mathematics in recent times. The need for an elementary book of this character is clearly evident from the changes which have occurred during the past twenty years, and which have been due, in part, to modifications in courses in pure mathematics and, in part, to new developments in applied mathematics.

The pure mathematics in this book includes essentially new developments, such as the theory of games, which has been applied widely to many problems ranging from business administration to industrial competition. It also includes other topics which, while not new to mathematicians, have lately come to play a greater part in the teaching and use of basic mathematics. These include set theory and its applications, number theory, and the transformations of coordinate systems and their geometric significance.

The applied mathematics in this book has been prepared to put at the command of the user of this book many of the modern methods used in business administration, notably in problems of maximinization of profits and production and minimization of costs and losses. These techniques include the transportation problem, linear and integer programming, network flow problems and combinatorial mathematics generally. Transforms are treated in some detail, and the important use of Laplace transforms in engineering and industrial control problems is explained. An entire chapter on probability and statistics culminates in an up-to-date treatment of quality control.

This book is designed for those who wish to gain such facility in understanding and using these newer methods by home study without the aid of a teacher. It can also be used to advantage by those who wish to broaden their knowledge of pure mathematics. It is one of a series of books on mathematics for home study, each volume of which is designed to assist those who wish to improve their ability to handle mathematical methods and operations.

DAVID B. MACNEIL

Brookside, New Jersey
August, 1963

v

CONTENTS

CONTENTS

CHAPTER 3 PROBABILITY, STATISTICS, AND QUALITY CONTROL

CHAPTER 4   THE THEORY OF GAMES

CHAPTER 5   INEQUALITIES, LINEAR PROGRAMMING, AND THE TRANSPORTATION PROBLEM

CONTENTS   ix

Chapter 1

THEORY OF SETS, NUMBERS, AND GROUPS

One of the most striking developments in modern mathematics has been the growth in importance of set theory. The reason for this development has been the fact that the entities or objects that are treated in every, or nearly every, branch of mathematics may be considered to be particular sets of entities or objects, so that their properties may be developed from general axioms of set theory. While to derive every branch of mathematics from set theory would be an extremely complex undertaking, there are sufficient basic applications of set theory, even at the level of this book, to justify its presentation in the opening chapter.

The concept of a set is intuitive, and is interchangeable with a class or a collection. The objects, or using the mathematical terms, the members or elements of the set may be of any sort. Thus there is the set of all birds, or the set of all Swedes. Mathematicians are more interested in such sets as the set of all prime numbers, or the set of all real numbers, or the set of regular polygons, or the set composed of −1, 0, and 1, or other sets of mathematical objects.

The procedure in the development of set theory is axiomatic, and it makes use of a distinctive symbolism. The meanings of the symbols are carefully defined. The entire structure of set theory is based on strictly logical inferences from the axioms, and requires some care at the start in learning the symbols and axioms. For that reason the number of symbols used in this chapter has been kept to a minimum, by stating in words those relations which occur less frequently.

1. Set Membership. The basic concept of set theory is that of set membership. If x is a member of the set A, this fact is expressed as

$$x \in A \tag{1-1}$$

Here set membership is expressed by the symbol $\in$, while the lower case letter x is used for the member and the capital letter A is used for the set.   This use of lower case and capital letters cannot always be followed, however, since the members of a set may themselves be sets, which when so considered, become subsets.

2. Equality.   The second concept, important here as in all mathematics, is that of equality.   The relation of equality between set A and set B which is expressed as

$$A = B \tag{1-2}$$

is defined by the Axiom of Extentionality, as follows:

(1) Two sets are equal if and only if they have the same members, i.e., sets A and B are equal if and only if every member of set A is a member of set B, and every member of set B is a member of set A.

In order to determine whether or not two sets are equal, we must know their members.   The members of a set may be expressed in two ways: by listing them or by describing them.   For example, consider the set D, which has as its members a, b, c, and d.   We can list the set D as

$$\{a, b, c, d\} \tag{1-3}$$

or we can describe it as

$$\{a \in D : a \text{ is the first or second or third or fourth letter of the alphabet}\} \tag{1-4}$$

The braces are commonly used to enclose the list or the description. The description reads "a is a member of (set) D: a being specified as one or another of the first four letters of the alphabet."

When one uses this second form of designating set membership, set description, he is said to specify the condition under which the elements of the set are determined.   This method leads logically to the Axiom of Specification.   Using the letter D for the four-member set in the above example, and the letter A for the set of all the letters of the alphabet, the Axiom of Specification would be stated as:

(2) To every set A and every condition $S(x)$, there corresponds a set D having as members exactly those members x of A for which $S(x)$ holds.

In the application, $S(x)$ is the condition that the members of set D be just those members of A (A being the set of letters of the alphabet) for which $S(x)$ holds, $S(x)$ being the specification that x be one or another of the first four letters of the alphabet.

The foregoing would be written as:

$$D = \{x \in A: \quad S(x)\} \tag{1-5}$$

3. Sets, Subsets and Inclusion. In the foregoing example, the set D of the first four letters of the alphabet, was a subset of the set A of all the letters of the alphabet, because A includes all the members of D. This relationship is symbolized as:

$$A \supset D$$

or $\qquad\qquad\qquad\qquad\qquad\qquad\qquad\qquad\qquad\qquad\qquad$ (1-6)

$$D \subset A,$$

the open end of the symbol being turned away from the subset. Note that the set in which the subset is included may not contain any other members than those in the subset; in other words the subset may be equal to the set. If it is not, that is, if the set has members that do not belong to the subset, the latter is called a *proper subset*.

To gain an idea of the number and variety of the subsets that can be found, consider the set D cited above, where

$$D = \{a, b, c, d\} \tag{1-7}$$

Now first of all D contains the equal subset $\{a, b, c, d\}$, which as defined above, is not a proper subset. D also contains a number of proper subsets. The three-membered subsets are $\{a, b, c\}$, $\{a, b, d\}$, $\{a, c, d\}$ and $\{b, c, d\}$. The two-membered sets, or pairs are $\{a, b\}$, $\{a, c\}$, $\{a, d\}$, $\{b, c\}$, $\{b, d\}$, and $\{c, d\}$. The unit sets are $\{a\}$, $\{b\}$, $\{c\}$, and $\{d\}$. Finally, there is an *empty subset*, defined as a set containing no members. This empty set is an important part of set theory, which considers that every set has one empty subset. It is symbolized by ϕ. Regarding it as a subset of above set $D = \{a, b, c, d\}$, it is found that set D has sixteen subsets in all, the identical subset, the four three-membered subsets, the six subset pairs, the four unit subsets, and the empty subset.

Similar writing of subsets for sets having one, two, three, five, etc. members discloses that the general formula for the number of subsets of a set is 2^n, where n is the number of members of the set.

In the above listing of subsets, it is to be noted that in no case was a subset repeated in changed order. In other words, sets (and, of course, subsets) are considered to be unordered unless, as will be explained later in this chapter, they are described and/or symbolized as being ordered.   If this is not done, then the set $\{a, b, c\}$ is equal to the set $\{c, b, a\}$, or to any other permutation.   Likewise, repetition of a member leaves the set unchanged, so that $\{a, b, c, a\} = A = \{a, b, c\}$.

4. Sets of Sets.   The concept of sets of sets can be approached by returning to the above discussion of the empty set ϕ, and by noting that the brackets were not placed around it as for the other sets that were found to be subsets of D.   For since ϕ is an empty set, it has no members.   On the other hand $\{\phi\}$ is the set containing the empty set ϕ as its only member.   Similarly $\{\{\phi\}\}$ is the set containing as its only member the set containing the empty set ϕ as its only member.   Thus

$$\{\{\phi\}\} \neq \{\phi\} \neq \phi, \text{ and } \{\{a\}\} \neq \{a\} \neq a.$$

Here the symbol $\neq$ means "not equal to" as explained in the following section.

5. Negatives.   The three relations which have been introduced up to this point have negatives.

Set membership which was symbolized by $\in$, has the negative symbol $\notin$, so that the statement that x is not a member of set A is expressed by $x \notin A$.   Equality of sets, which was symbolized by $=$, has the negative symbol $\neq$, so that the statement that A is not equal to B, is expressed by $A \neq B$.   Inclusion, which was symbolized by $A \subset B$ or $B \supset A$, has the negative $\not\subset$ or $\not\supset$, so that the statement that A is not included in B is expressed by $A \not\subset B$ or $B \not\supset A$.

6. Unions and Intersections.   As with numbers and other mathematical objects, there are operations which may be performed upon two or more sets, yielding still other sets.

The *union* of two sets is defined as the set containing all the elements that belong to at least one of the sets.   Thus if set C is the union of sets A and B, it is written

$$C = A \cup B \qquad (1\text{-}8)$$

and is shown diagramatically in Figure 1-1.   (Diagrams of this type used to demonstrate or investigate set theoretic and other logical propositions are called Venn diagrams.)

In this figure the larger circle defines set A by enclosing all of its elements, the smaller circle defines set B by enclosing all its elements, and the entire shaded area defines set C which is the union of sets A and B.

The concept of union of sets is not restricted to two sets but may be extended to any number of sets by the Axiom of Unions:

For every collection of sets there is a set that contains all elements belonging to one or more of the sets of the collection. Note that this axiom does not state that the inclusive set

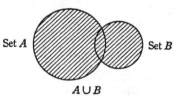

FIG. 1-1. The union (shaded) of set A and set B.

may not contain elements not found in any of the sets of the collection. To impose this restriction on the union, the following definition is used.

Denoting the collection of sets by **C**, the union can be defined in the form: For every collection of sets **C** there exists a set D such that if (and only if) x is a member of X, where X is any set in **C**, then x is a member of D; this definition being is written symbolically as

$$\{x \in D : x \in X \text{ for any } X \text{ in } \mathbf{C}\} \tag{1-9}$$

This axiom also suggests a briefer expression for the union D, namely

$$D = \bigcup \{X : X \in \mathbf{C}\} \tag{1-10}$$

which designates D as the union of the sets in the collection **C**.

The *intersection* of two sets A and B is defined as the set containing only those elements that are in A and in B. Thus if the set C is the intersection of sets A and B, it is written

$$C = A \cap B \tag{1-11}$$

and is shown diagrammatically in Figure 1-2.

In this figure, as in Figure 1-1, the larger circle represents set A, the smaller, set B, and the shaded area, the intersection of A and B. The concept of the intersection of sets is not limited to two sets, but may be extended to a collection of sets. Thus, Fig. 1-3 shows the intersection of sets A, B and C.

In this figure, the three circles represent the three sets, the lightly shaded area represents the intersection of A and B, and the heavily shaded area represents the intersection of set C with the intersection of A and B.

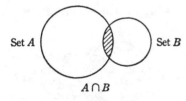

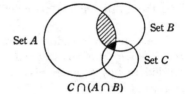

$A \cap B$

$C \cap (A \cap B)$

FIG. 1-2. The intersection (shaded) of A and B.

FIG. 1-3. The intersection (solid shading) of C with the intersection (shaded) of A and B.

The general formulation for the intersection of a collection of sets $\mathbf{C}$ is

$$D = \{x : x \in X \text{ for every } X \text{ in } \mathbf{C}\} \tag{1-12}$$

where D is the intersection. D may also be denoted by

$$D = \bigcap (X : X \in \mathbf{C}) \tag{1-13}$$

Note that in the three-set intersection, the expression obtained contained two operational symbols, both of which were intersections. Set theory is concerned quite generally with expressions or relations involving more than one symbol, and their interrelations constitute one of the most important parts of the subject. Some of the more general are discussed in the next paragraph.

7. Relations Involving Unions and Intersections.

1. Both union and intersection of sets are commutative, that is

$$A \cup B = B \cup A \tag{1-14}$$

$$A \cap B = B \cap A \tag{1-15}$$

These two relations follow from the definition of union and the concept of set membership. However, throughout this section formal proofs are not given.

2. Both union and intersection of sets are associative, that is

$$(A \cup B) \cup C = A \cup (B \cup C) = A \cup B \cup C \tag{1-16}$$

$$(A \cap B) \cap C = A \cap (B \cap C) = A \cap B \cap C \tag{1-17}$$

3. Intersection distributes over union, that is

$$A \cap (B \cup C) = (A \cap B) \cup (A \cap C) \tag{1-18}$$

The intersection of A and the union of (B and C) equals the union of (the intersection of A and B) and (the intersection of A and C). (See Fig. 1-4).

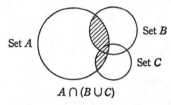

$$A \cap (B \cup C)$$

Fig. 1-4. The intersection (shaded) of A and the union of (B and C).

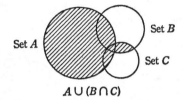

$$A \cup (B \cap C)$$

Fig. 1-5. The union (shaded) of A and the intersection of (B and C).

4. Union distributes over intersection, that is

$$A \cup (B \cap C) = (A \cup B) \cap (A \cup C)$$

The union of A and the intersection of (B and C) equals the intersection of (the union of A and B) and (the union of A and C). (See Fig. 1-5).

8. The Complement. Further discussion of set relationships requires the introduction of the complement. If A and B are sets, the relative complement of B in A, also called the difference of A and B, and expressed as $A - B$, consists of those members of A which are not present in B, as shown in Figure 1-6.

By the use of the complement, two important set theoretic relationships, known as de Morgan's laws, may be expressed as

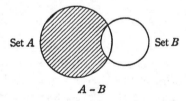

$$A - B$$

Fig. 1-6. The relative complement (shaded) of B in A.

(a) $(A - B) \cap (A - C) = A - (B \cup C)$ (1-19)

(b) $(A - B) \cup (A - C) = A - (B \cap C)$ (1-20)

Law (a) may be stated in words as: The intersection of the complements of B in A and C in A equals the complement of (the union of B and C) in A. (Fig. 1-7).

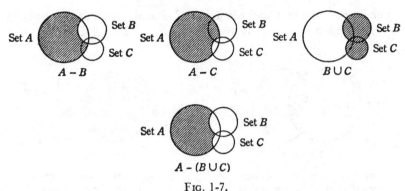

Fig. 1-7.

Law (b) may be stated in words as: The union of the complements of B in A and C in A equals the complement of (the intersection of B and C) in A.   (Fig. 1-8).

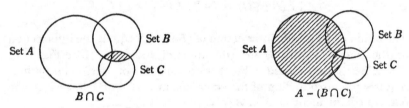

Fig. 1-8.

The de Morgan laws are sometimes called *dualization laws* because one is changed into the other by reversal of all inclusions and replacement of the intersection symbol by that for union, and the symbol for union by the symbol for intersection.   Thus, one law may be derived from the other.   In fact, many of the theorems of set theory behave similarly, and are accordingly easier to induce and remember.

9. The Absolute Complement.   If A is a set, its absolute complement may be defined as $E - A$, where E is the *universal set*.   We can then write $E - A = A'$, where A' is the set of all members of the universal set which are not members of A.   Then the de Morgan laws can be written, for the two sets A and B, as

$$(A \cup B)' = A' \cap B' \tag{1-21}$$

$$(A \cap B)' = A' \cup B' \tag{1-22}$$

These statements of de Morgan's laws are useful in clearing parentheses when working with sets.

Other useful expressions involving absolute complements are

(1) $$(A')' = A,$$ (1-23)

the absolute complement of the absolute complement of A is A.

(2) $$A \cap A' = \phi$$ (1-24)

the intersection of a set with its absolute complement is the empty set.

(3) $$A \cup A' = E$$ (1-25)

the union of a set with its absolute complement is the universal set.

(4) $$E' = \phi \quad \phi' = E,$$ (1-26)

the absolute complement of the universal set is the empty set, and the absolute complement of the empty set is the universal set.

10. Powers. In set theory, the power set of a set is that set which contains all the subsets of the given set. The Axiom of Powers asserts the existence of such a power set (collection of sets). The definition can be expressed as

$$\mathbf{P}(A) = \{X : X \subset A\},$$ (1-27)

which may be stated in words as: The power set $\mathbf{P}(A)$ is composed of every subset X of set A. For example, the power set $\mathbf{P}(D)$ of set D which was discussed on page 3 would contain all sixteen of the subsets of D which were enumerated.

It would thus be expressed as

$$\mathbf{P}(D) = \mathbf{P}\{a, b, c, d\}$$

$$= \{\{ a, b, c, d\}, \{a, b, c\}, \{a, b, d\}, \{a, c, d\}, \{b, c, d\}, \{a, b\}, \{a, c\},$$
$$\{a, d\}, \{b, c\}, \{b, d\}, \{c, d\}, \{a\}, \{b\}, \{c\}, \{d\}, \{\phi\}\} \quad (1\text{-}28)$$

As stated earlier, a set of n members has 2^n possible subsets; therefore a power set of an n-member set has 2^n members.

11. Ordered Parts and Cartesian Products. It was explained on Page 3 that the general definition of a set makes no provision for ordering. The expression $A = \{a, b, c, d\}$ merely states that the members of set A are a, b, c, d (and, of course, the empty set ϕ). Therefore $A = \{a, b, c, d\} = \{d, c, b, a\} =$ any other bracketed permutation of the four letters.

However, set theory provides a means for symbolizing and describing ordered sets. Consider the ordered pair of a and b, with first coordinate a and second coordinate b. The order of this pair may be expressed by defining it as a set consisting of the two subsets $\{a\}$ and $\{a, b\}$, which is written as $(a, b) = \{\{a\}, \{a, b\}\}$.

It is important at this point to note a significant difference between sets of two members and ordered pairs. The set $\{a, b\}$ is equal to the set $\{b, a\}$; the ordered pair (a, b) is by no means necessarily equal to the ordered pair (b, a).

The concept of ordering can be extended in several directions from the ordered pair. The ordered triple $(a, b, c) = \{\{a\}, \{a, b\}, \{a, b, c\}\}$ is a logical extension of the ordered pair, and is useful in number set theory for describing the position of a point, or a vector, in three-dimensional space. In fact, the ordered n-ple is a general means of describing the position of a point in n-dimensional space.

Another extension of the concept of the ordered pair is the set of ordered pairs. For if A and B are sets, and if a is a member of A and b is a member of B, it can be shown that the ordered set (a, b) is a subset of $\mathbf{P}(A \cup B)$, and that therefore it is an element of $\mathbf{P}(\mathbf{P}(A \cup B))$. Moreover, it can also be shown that a set exists which consists only of the ordered pairs (a, b) where a is a member of A and b is a member of B. This set of ordered pairs is named the Cartesian product of A and B, and shown symbolically as $A \times B$.

The converse proposition can also be shown to be true. If the set C has only ordered pairs as its members, it can be shown that C is the Cartesian product of two sets A and B such that each member of A designated as a corresponds to some b, both forming the ordered pair (a, b) that is a member of C, and also that each member of B, designated as b, corresponds to some a, both forming the ordered pair (a, b) that belongs to C. These sets A and B are called the projections of C on to the respective coordinates.

12. Relations: Function, Domain and Range. One of the most important contributions of set theory to mathematical logic (as well as to other branches of mathematics) is the use of the ordered pair to express binary relationship. Representative binary relationships are husband and wife or father and son. Since a pair consisting of one husband and one wife is necessary for the first relationship to exist and a pair consisting of one son and one father for the second to exist, and since the binary relationship is necessary for the unique pair to exist, then the binary relationship and the unique pair are reciprocal entities, as was found for the ordered pair and the Cartesian product. The ordered pair formulation of the marriage and father-son relationships can therefore be written as

$$(h, w) \in M \qquad (1\text{-}29)$$

$$(s, f) \in P \qquad (1\text{-}30)$$

where M is the set of married couples, or the *relation of marriage*, and P is the set of sons and fathers, or the relation of paternity to male offspring.

There are several kinds of relations that are given class names in set theory. A reflexive relation applies to the elements of one set only, and is symbolized as aRa for every member a of A. An example would be measurement of the same quantity (i.e. the weight of an object) in different systems of units, e.g. 906 grams R 2 pounds. A symmetric relation is one for which aRb implies bRa (written $aRb \rightarrow bRa$); marriage is symmetric. A relation is transitive if aRb and bRc imply that aRc (written aRb, $bRc \rightarrow aRc$). An example is: a Prussian is a German and a German is a European, implies that a Prussian is a European.

A *function* is defined as the relation between a certain object from one set, called the *range*, with each object from another set, called the *domain*. For example, a function might be defined as having as its value a person's age when the person is specified — it would then be said that a person's age is a function of the person, and that the domain of this function is the set of all human beings and the range is the set of all integers which are the ages of all living persons.

13. Functions.   A function is a restricted type of relationship. The restriction limits the ordered pairs that are its members to those of which no two have the same first coordinate. In terms of the definitions of relation and domain, a function from the *set X to the set Y is a relation f such the domain of f is X, and for each member x of X there is a unique member in Y with the ordered pair (x, y) a member of f. To illustrate the difference between a relation and a function, the relation son to father meets the requirement of a function, since there is only one father for each son. However, the relation father-son does not, since one father may have more than one son.

At this point, it is necessary to explain a difference between the terminology of set theory and a long-established and still current mathematical usage. When one speaks of single-valued functions and multiple-valued functions, he is using the word "function" in the sense of the term "relation" of set theory, rather than that of the term "function." As stated above, and as will be illustrated below, in set theory a function is single-valued by definition, that is, in the function $y = f(x)$, there is only one value of y corresponding to a particular value of x. In this sense, $y^2 = x$ is a relation, and not a function.

In set-theoretic formalism, if $(x, y) \in f$ and $(x, z) \in f$, then $y = z$. In words, if we imagine two members in y which form ordered pairs with one member in X, then the two must be equal, for they cannot be different. Note that the definition as italicized states that the domain of f is X, but makes no statement about the range of f, which need not be equal to Y. (If the range of X is equal to Y, then f is said to map X onto Y.) The range of f is simply the set of second coordinates of the members of f (ordered pairs).

Since the function is one of the most useful concepts in set theory, it is illustrated by several examples of number sets.

Consider the set $\{(1, 2), (3, 4), (5, 6), (7, 8)\}$. It is obviously a function, since it meets all the criteria given in the definition.

The set $\{(1, 2), (3, 4), (5, 2), (6, 4)\}$ is a function for the same reason. Note that in two cases the same member in Y is paired with two different members in X, that is, 2 occurs in $(1, 2)$ and $(5, 2)$ and, 4 occurs in $(3, 4)$ and $(6, 4)$. However, the definition does not exclude this repetition of members in Y.

The set $\{(1, 2), (3, 4), (5, 6), (5, 7)\}$ is not a function because the last two ordered pairs have the same member in X, which is 5, for two different members in Y; therefore they do not have a unique member in Y as required by the definition.

The criteria can also be applied to graphical representations of sets as shown in Fig. 1-9.

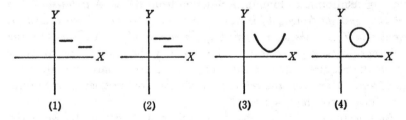

Fig. 1-9.

The graphs in Fig. 1-9 are considered to represent series of ordered pairs consisting of members in X and members in Y that are close in numerical magnitude so that any point on the graphs may be treated as representing an ordered pair belonging to the set of ordered pairs. It can be seen that (1) and (3) are functions, since each member in X has a unique companion in Y. On the other hand, (2) and (4) are not functions since in both there are members in X which have two corresponding members in Y.

In the case of number sets which are defined by formula, it is often necessary to describe the domain and range. Thus, the function $x^3 + x^2 + 2x + 10$ has as its domain the set of all numbers, and as its range the values found by substituting each number in $x^3 + x^2 + 2x + 10$. This function is therefore expressed as

$$f(x) = x^3 + x^2 + 2x + 10.$$

The use of this symbol $f(x)$ recalls to mind the fact that this notation is often written with more than one letter in parentheses. For example, the expression $x^3 + 2x^2 + 5xy^2 + 2y^3 + 10$ is designated as $f(x, y)$. The natural question is whether such expressions are functions and how they can be treated within the framework of set theory. The answer is that the treatment of functions given in this chapter has been based on the ordered pair. However, by starting from the ordered triple (See page 10) a definition of function can be developed to take care of the three-set situation. Rather than proceeding with this extension, which would add nothing essentially new in principle, and rather than treating certain other phases of set theory, such as partial ordering, we will proceed now to the treatment of the numbers themselves.

14. Number Sets. The difficulty in defining numbers arises from the necessity of establishing other concepts in terms of which such definitions may be formulated. Set theory provides such concepts. While rigorous formulation of them requires the derivation of several other concepts in set theory in addition to those already defined in this chapter, we can proceed at this point to develop a definition of cardinal numbers in an intelligible, if not wholly rigorous, fashion.

In the preceding section, a function from the set X to the set Y was defined to mean, in part, that for each member in X there is a unique member in Y. Now if there is also a function from the set Y to the set X, then for each member in Y there is a unique member in X. That is, there is a one-to-one correspondence between X and Y and Y and X. This symmetric relationship is expressed by stating that sets X and Y are similar, which is symbolized as

$$X \approx Y \tag{1-31}$$

The cardinal number of a set can then be defined as the class of all sets similar to it, symbolized as

$$[\mathbf{n}(X) = \mathbf{n}(Y)] \text{ implies } X \approx Y \tag{1-32}$$

$$X \approx Y \text{ implies } [\mathbf{n}(X) = \mathbf{n}(Y)]$$

or in accordance with the concepts defined earlier in this chapter,

$$\text{dom } \mathbf{n} = \mathbf{C}(S) \tag{1-33}$$

$$\text{ran } \mathbf{n} = T[\mathbf{C}(X, Y)] \tag{1-34}$$

The domain of $\mathbf{n}$ is the class of all sets considered, and the range of $\mathbf{n}$ is any set having as its members classes of similar sets.

Since one-to-one correspondence is transitive, then if

$$A \approx B \text{ and } B \approx C, \text{ then } A \approx C.$$

Therefore, from the above relations, *every* set has a unique cardinal number $\mathbf{n}$ associated with it. Moreover, since the class of all sets similar to any one of them have the same cardinal number, we can define that cardinal number in terms of the elements in each of the similar sets.

Thus, 0 is defined as the cardinal number of the empty set, 1 is defined as the cardinal number of the singleton, 2 is defined as the cardinal number of the pair $\{0, 1\}$, 3 is defined as the cardinal number of the triple $\{0, 1, 2\}$, and so on, or symbolically

$$\left. \begin{array}{l} 0 = \mathbf{n}(\phi) \\ 1 = \mathbf{n}(\{0\}) \\ 2 = \mathbf{n}(\{0, 1\}) \\ 3 = \mathbf{n}(\{0, 1, 2\}) \end{array} \right\} \tag{1-35}$$

and so on.

The foregoing is a form of statement of the first two of the Peano axioms, which assert that the set of all natural numbers, N is the unique successor set that is a subset of every successor of it

$$(0 = \phi) \in N$$

if

$$n \in N \text{ then } n \cup \{n\} \in N \qquad (1\text{-}36)$$

The third of the Peano axioms is called the Principle of Mathematical Induction. It generalizes the succession of equalities given above to define the natural numbers by stating that if any set A is contained in the set of natural numbers N, and if 0 is a member of A and if the union of cardinal n and set $\{n\}$ (which is $n + 1$) is a member of A whenever n is a member of A, then the set A is the set of natural numbers N.

The fourth Peano axiom merely asserts that no successor set is equal to zero, while the fifth axiom asserts that if two successor sets n and m are members of N, the set of all natural numbers, and if $n \cup \{n\} = m \cup \{m\}$, then $n = m$.

The Peano axioms and the methods of set theory furnish a basis for the rigorous derivation of the properties of natural numbers as well as other classes of numbers. However, to do so in this book would require so many pages that many other topics must be omitted. Therefore, the remainder of the treatment of numbers will be devoted to the properties themselves, with a minimum of proof.

From the Peano axioms it can be deduced that the set of all natural numbers N is infinite, since by the first axiom if a number n, however great, is a member of N, then $n + 1$ is a member of N. However, while infinite, N is countable (or denumerable) because it can be proved that the members of N can be written as a sequence 0, 1, 2, 3, 4 $\cdots$.

15. Operations with the Natural Numbers. In order to use any class of mathematical objects, it is necessary to define what relations may be established between them, or in other words, what operations may be performed on them. In the case of the natural numbers 0, 1, 2, 3, 4 $\cdots$ both addition and multiplication may be performed, because the addition of any two natural numbers (members of the set) produces a natural number (member of the set). Also, the multiplication of any two natural numbers yields a natural number. The set of natural numbers is thus said to be closed under addition and multi-

plication.   It is not closed under subtraction or division, as is proved
by the operations $1 - 2$ and $1 \div 2$, which do not produce natural
numbers.

16. Integral Numbers or Integers.   When we extend the class of
natural numbers by including their negatives, we obtain the integers,
defined as $\cdots -4, -3, -2, -1, 0, 1, 2, 3, 4 \cdots$.   Now we find that
this set is closed not only under addition and multiplication, but also
under subtraction, as illustrated by the operation $1 - 2$, which now
produces a member of the set, -1.   The set of integers is not closed
under division, as is shown by the fact that $1 \div 2$ does not yield a
member of the set.

It is to be noted that the subsets of an infinite countable set may
themselves be infinite, and may be closed under some, or all, of the
operations under which the set is closed.   Thus, the set of odd integers
is closed under multiplication, but not under addition or subtraction
(or, of course, division) while the set of even integers, like that of all
integers, is closed under addition, subtraction and multiplication.

17. Rational Numbers.   A convenient way in which to fix in mind
the nature of rational numbers is to investigate the numbers that must
be added to the set of integers so that it is closed to division.   This can
be done by adding to the set of integers $\cdots -4, -3, -2, -1, 0, 1, 2,$
$3, 4 \cdots$, all fractions expressible by choosing any member of the set as
its numerator and any other member as its denominator, provided that
0 in the denominator is excluded.   The reason for excluding division
by 0 is not only that we cannot write a number for the quotient that
can be written as part of a sequence, but also that the reverse operation,
i.e., multiplication by 0, does not yield a unique product.

Fractional rational numbers are not always stated as ratios in their
lowest terms, but in some cases as decimal numbers, e.g., .5164.
Therefore, it is convenient to establish working methods for determining whether decimal numbers are rational numbers.

We can begin by recalling that every terminating decimal is also a
fraction having a denominator that is a power of ten.   Thus, .2 is
$2/10$, 45 is $45/10^2 = 45/100$, .254 is $254/10^3 = 254/1000$.   Therefore,
it follows at once that every terminating decimal is a rational number.

Now let us consider the periodic decimals, taking as an example
$.588\overline{235288}$ where the bar over the last six figures indicates that the
decimal does not terminate, but continues to repeat the six figures,
235288; that is, the value of this decimal, if carried out until the repeat-

ing portion appeared twice, would be .588235288$\overline{235288}$. Now let us multiply both the one period and the two period forms by powers of 10 that will make both of them whole numbers. The second form has 15 decimal places, so we multiply it by 10^{15}; the first one has nine decimal places, so we multiply by 10^9. We then have on subtracting the products (representing the given decimal by D):

$$D \times 10^{15} = 588,235,288,235,288 + \text{periodic continuation}$$
$$D \times 10^9 = 588,235,288 + \text{periodic continuation}$$

$$D \times (10^{15} - 10^9) = 588,234,700,000,000$$

or

$$D = \frac{588,234,700,000,000}{10^{15} - 10^9}$$
$$= \frac{588,234,700,000,000}{999,999,000,000,000}.$$
$$= \frac{5,882,347}{9,999,990}$$

which is a rational number. Therefore we can infer that every periodic decimal is a rational number.

18. Real Numbers. The last part of the above section on rational numbers showed that not only the fractions of the type a/b, where a and b are positive or negative integers, were rational numbers, but also that terminating and periodic decimals are rational numbers. There remains, of course, the entire class of decimal numbers which are neither terminating nor periodic. While such numbers are not rational, they are real, as is shown by the fact that they describe the position of a point in a one-dimensional continuum. In fact, the latter criterion is an effective definition of the set of real numbers. It includes both the set of rational numbers, and the set of real numbers which are not rational, and which are therefore called the irrational numbers.

One class of irrational numbers is that of algebraic irrational numbers. *Algebraic numbers* are defined as numbers which satisfy algebraic equations of the form:

$$a_n x^n + a_{n-1} x^{n-1} + \cdots + a_1 x + a_0 = 0,$$

and may, therefore, be rational or irrational.

The algebraic irrational numbers include all the roots of positive numbers which are not corresponding powers of rational numbers. Thus, $\sqrt{2}$, $\sqrt{3}$, $\sqrt{5}$, $\sqrt{6}$, $\sqrt{7}$, $\sqrt{8}$, $\sqrt{10}$ and so on (skipping $\sqrt{16}$, $\sqrt{25}$ and the other square roots of binary powers of integers) are algebraic irrational numbers.   So are $\sqrt[3]{2}$, $\sqrt[3]{3}$, $\sqrt[3]{4}$, $\sqrt[3]{5}$, $\sqrt[3]{6}$, $\sqrt[3]{7}$, $\sqrt[3]{9}$, and so on, skipping $\sqrt[3]{8}$, $\sqrt[3]{27}$, $\sqrt[3]{64}$, and the other cube roots of third powers of integers.   The algebraic numbers obviously extend to all the higher roots indefinitely.   Moreover, they include very many of the sums, differences, products and quotients of their class, *but not all.* In other words, the class of algebraic irrational numbers is not closed under addition (subtraction), multiplication or division.   This is evident at once from $\sqrt{2} - \sqrt{2} = 0$, $\sqrt{2} \cdot \sqrt{2} = 2$, and $\sqrt{8}/\sqrt{2} = 2$. However, it is readily apparent that any of these operations performed upon irrational numbers taken at random are far more likely to produce irrational than rational numbers.

All numbers obtained by combining an irrational number with one or more rational numbers, whether the combination operation is addition (subtraction), multiplication or division, or more than one such operation, produces an irrational number.

It was stated earlier that every algebraic number is the root of an algebraic equation.   From this definition it follows that algebraic numbers included all rational numbers as well as certain irrational ones.   For the equation $x^2 - 7x + 12 = 0$, is clearly an algebraic equation, and it has as its roots $x = 3$ and $x = 4$, which are plainly rational numbers.

19. Complex Numbers.   The discussion of algebraic numbers, and numbers in general, up to this point has been restricted to the real numbers, rational and irrational.   However, any algebraic equation of the form $x^2 + a = 0$, where a is a positive rational number, has the roots $x = \pm\sqrt{-a}$, and such numbers are called *imaginary numbers.*   In many operations, it is more convenient to write them in the form $x = \pm\sqrt{a}\sqrt{-1}$ or $x = \pm\sqrt{a}i$, where i is the symbol of $\sqrt{-1}$.   This notation is especially useful in operations with *complex numbers,* which are ordered pairs consisting of one real number and one imaginary number.

Addition and Subtraction of Complex Numbers.   Let $a + ib$ and $c + id$ represent two complex numbers.   The sum of the two is then

$$(a + ib) + (c + id) = a + c + ib + id,$$

or, taking out the common factor i in the last two terms,

$$(a + ib) + (c + id) = (a + c) + i(b + d).$$

Similarly,

$$(a + ib) - (c + id) = (a - c) + i(b - d).$$

$$(1\text{-}37)$$

The sum (or difference) of two complex numbers is, therefore, another complex number whose real part is the sum (or difference) of the real parts of the two numbers and whose imaginary part is the sum (or difference) of their imaginary parts.

Since a complex number is composed of two parts which are always distinct, it cannot be zero unless both parts are separately equal to zero. This is an important property of complex numbers, which we now make use of to show another important property of these numbers.

Suppose two complex numbers are equal; then the difference between them is zero. That is, if $a + ib$ and $c + id$ are two complex numbers, and if

$$a + ib = c + id,$$

then

$$(a + ib) - (c + id) = 0.$$

But, according to (1-37) this difference is $(a - c) + i(b - d)$. Therefore,

$$(a - c) + i(b - d) = 0.$$

This is a complex number whose real part is $a - c$ and whose imaginary part is $b - d$, and it is equal to zero. But for this complex number to be equal to zero its real and imaginary parts are each equal to zero, as seen above. Therefore,

$$a - c = 0, \quad \text{and} \quad b - d = 0.$$

But, when the difference between two numbers is zero, the two are equal. Therefore,

$$a = c, \quad \text{and} \quad b = d.$$

Now a, c and b, d are the real and imaginary parts of the two original equal complex numbers, and they are, respectively, equal. We therefore, have the important result that

If two complex numbers are equal their real parts must be equal and their imaginary parts equal.

This result may be expressed in symbols by saying that if

$$A = a + ib, \quad B = c + id,$$

then if

$$A = B, \quad a = c \quad \text{and} \quad b = d.$$

$$(1\text{-}38)$$

If we use the same form of expression for the results (1-37) as is used above for (1-38) we can say that: If

$$A = a + ib, \quad B = c + id,$$
$$A = B = x + iy,$$

where

$$x = a + c, \quad y = b + d.$$

$$(1\text{-}39)$$

Multiplication and Division of Complex Numbers. The product of two complex numbers $a + ib$ and $c + id$ is found in the same way as that of any two binomials. Thus:

$$\begin{array}{r} a + ib \\ c + id \\ \hline iad + i^2bd \\ ac + ibc \\ \hline ac + i(ad + bc) + i^2bd \end{array}$$

But $i^2 = -1$ and, therefore, $i^2bd = -bd$. The product is, therefore, $ac + i(ad + bc) - bd$, or

$$(a + ib)(c + id) = (ac - bd) + i(ad + bc).$$

This is also a complex number whose real part is $ac - bd$ and whose imaginary part is $ad + bc$. The product of any two complex numbers, is, therefore, another complex number whose real part is the difference between the product of the real and that of the imaginary parts of the factors, and whose imaginary part is the sum of the cross products of the real and imaginary parts of the factors.

This result is expressed in symbols by saying that, if

$$A = a + ib, \quad B = c + id,$$
$$AB = x + iy,$$

where

$$x = ac - bd, \quad y = ad + bc.$$

$$(1\text{-}40)$$

The square of the complex number $a + ib$ is written in the usual way and has the same meaning as the square of any binomial, that is,

$$(a + ib)^2 = (a + ib) \cdot (a + ib).$$

If we apply the multiplication rule (1-40) to this product we find that,

$$(a + ib)(a + ib) = (a^2 - b^2) + i(ab + ab),$$

or, $$(a + ib)^2 = (a^2 - b^2) + i \cdot 2ab.$$

In the same way we find that,

$$(a - ib)^2 = (a^2 - b^2) - i \cdot 2ab.$$

Both these results may be concisely expressed by letting

$$A = a \pm ib;$$

then $$A^2 = x \pm iy,$$ (1-41)

where $$x = a^2 - b^2, \quad y = 2ab.$$

The quotient of two complex numbers $A = a + ib$ and $B = c + id$ is written in the usual form

$$\frac{A}{B} = \frac{a + ib}{c + id}.$$

As seen before, this fraction may have its numerator and denominator multiplied by the same number without changing its value.   Let us multiply by $c - id$.   This gives

$$\frac{A}{B} = \frac{(a + ib)(c - id)}{(c + id)(c - id)}.$$

By the method given above, the product in the numerator of the fraction on the right is found to be $(ac + bd) + i(bc - ad)$ and furthermore, the product in the denominator (product of sum and difference of same numbers) is $c^2 - i^2 d^2$, or, since $i^2 = -1$, $c^2 + d^2$.   The quotient is, therefore,

$$\frac{A}{B} = \frac{(ac + bd) + i(bc - ad)}{c^2 + d^2}$$

that is,

$$\frac{A}{B} = \left(\frac{ac + bd}{c^2 + d^2}\right) + i\left(\frac{bc - ad}{c^2 + d^2}\right).$$

The quotient of two complex numbers is, therefore, also a complex number.

The result just obtained can be expressed by writing

and

where

$$A = a + ib, \quad B = c + id,$$
$$A/B = x + iy,$$
$$x = \frac{ac + bd}{c^2 + d^2}, \quad y = \frac{bc - ad}{c^2 + d^2}.$$

$$(1\text{-}42)$$

Absolute Value of a Complex Number. In finding an expression for the quotient of two complex numbers in the preceding paragraph, we made use of a product of the form

$$(a + ib)(a - ib) = a^2 + b^2.$$

Now, this result contains no imaginary part. Therefore, we have here a real number as the product of two complex numbers. This special case does not conform to the general rule for multiplication of complex numbers and is of particular significance.

It is to be noted that the two complex numbers $a + ib$ and $a - ib$ are closely related in a very simple manner: they differ only in the sign of their imaginary parts. Two complex numbers related in this manner are said to be conjugate complex numbers. The conjugate of the complex number A is represented by $\overline{A}$ (called "A bar"). Thus, the conjugate of the complex number

is

$$A = a + ib$$
$$\overline{A} = a - ib$$

$$(1\text{-}43)$$

and their product is

$$A\overline{A} = a^2 + b^2.$$

The square root of this product is also a real number and is called the absolute value of the complex number A. This is written $|A|$. Using these symbols and the values of A and $\overline{A}$ as given by (1-43)

$$|A|^2 = A\overline{A} = a^2 + b^2,$$
$$\therefore |A| = \sqrt{a^2 + b^2}.$$

$$(1\text{-}44)$$

Similarly, if $B = c + id$, the absolute value of B is $|B| = \sqrt{c^2 + d^2}$; if $A = 4 + 3i$, $|A| = \sqrt{4^2 + 3^2} = \sqrt{25} = 5$. While the absolute value, being a square root, may be written as either plus or minus, the positive value is always taken. Thus, when

$$A = 4 + 3i, \quad |A| = 5.$$

Two complex numbers may be said to be equal in absolute value, or one may be said to be greater or less than another in absolute value, in the usual sense of speaking of positive numbers.

Representation of Complex Numbers. A mapping of complex numbers onto a Cartesian plane, i.e. by the use of Cartesian coordinates, is called an Argand diagram (or plane) if the complex number $x + iy$ corresponds to the vector (or its end point) issuing from the origin with end point x, y. The X-axis is the real axis and the Y-axis is the imaginary axis. Thus the complex number $a + ib$ is plotted by measuring a units along the X-axis and b units along the Y-axis and projecting the corresponding ordinate and abscissa to their point of intersection. The distance to this point from the origin, which is thus equal to $\sqrt{a^2 + b^2}$ is the modulus or absolute value of the complex number (vector), as shown above in Eq. (1-44). By taking the plotted endpoint as a new origin and plotting another complex number from there in the same way, a new point is located which is the sum of the two complex numbers. Similarly, this representation in coordinate systems facilitates other operations with complex numbers besides addition (or subtraction).

For such calculations, representation in polar coordinates is often even more useful than in rectangular Cartesian ones. (For a review of polar coordinates, see Chapter 7). If for a complex number $a + ib$, r and θ are chosen so that $a = r \cos \theta$ and $b = r \sin \theta$, one obtains the complex number in the form $r(\sin \theta + i \cos \theta)$ which is convenient for use in calculation because of the properties of the trigonometric functions. Thus to raise a complex number to an integral power, one can use the de Moivre identity.

$$r(\cos \theta + i \sin \theta)^n = r^n(\cos n\theta + i \sin n\theta) \qquad (1\text{-}45)$$

Moreover, since $(\cos \theta + i \sin \theta) = e^{i\theta}$ (Euler's formula), then this power formula may also be written

$$(re^{i\theta})^n = r^n e^{in\theta} \qquad (1\text{-}46)$$

Multiplication of complex numbers is also facilitated in the polar form, for the product of two complex numbers, $re^{i\theta}$ and $r'e^{i\phi}$ is simply $rr'e^{i(\theta+\phi)}$.

20. Trigonometric and Logarithmic Numbers. Earlier in this discussion it was pointed out that any numbers which are not algebraic are called transcendental. The general discussion of transcendental numbers if given in the next section; now we deal with two important kinds of numbers, the trigonometric and logarithmic numbers, which with certain rational exceptions to be noted, are transcendental.

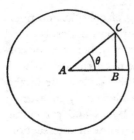

Since the trigonometric functions are the ratios of the sides of a right triangle, expressed for successive values of one of its acute angles, it is not difficult to picture why they are largely irrational. (This relation is readily seen from Figure 1-10, in which a right triangle ABC is drawn so that its hypotenuse AC is the radius of a circle, and its side AB lies on another radius. The trigonometric functions are, of course, the ratios of the sides of this triangle, and if AC is taken as unity obviously most lengths of AB and BC, obtained by rotation of AC, will be irrational.

Fig. 1-10. Values of trigonometric functions are the six ratios of AB, AC, and BC, for increasing values of θ resulting as AC rotates counterclockwise from AB.

Certain trigonometric functions (numbers) are rational, e.g., sin 0°, cos 0°, sin 30°, sin 90°, cos 90°, tan 45°. In general, the rational values are the functions of 0° and 90°, and in some, but not all, functions of 30°, 45° and 60° and any angle obtained from these by the operation $\pm n \cdot 90°$.

Just as the relationships of trigonometric functions to the motion of the radius vector of a circle suggests why such trigonometric numbers are irrational, so the relation of logarithms to powers and roots suggests why logarithms are irrational. Moreover, it also suggests why the irrationality is independent of the base of the system. The only logarithms that are rational are those of numbers that are integral powers (positive or negative) of the base of the system, *provided that the base is itself a rational number, such as 10.*

21. Transcendental Numbers. The conditions made at the end of the preceding paragraph for those logarithms that are rational, namely that they be integral powers of the base, and also that the base be itself rational, serve readily to introduce the subject of transcendental numbers. For the number e, the base of natural logarithms is a well-known transcendental number. It is represented by the infinite series

$$e = \lim_{x \to 0} (1 + x)^{1/x} = 1 + \frac{1}{1!} + \frac{1}{2!} + \frac{1}{3!} + \frac{1}{4!} \cdots \frac{1}{n!} \cdots$$

$$= 1 + 1 + 1/2 + 1/6 + 1/24 + \cdots + 1/n! \ldots \ldots \ldots$$

$$= 2.71828 \ldots \ldots \ldots \ldots \quad (1\text{-}47)$$

The reason it is called a transcendental number is because it is not an algebraic number, as was stated earlier in this chapter. That is, it is not the root of an algebraic equation. The proof of this statement is necessarily a negative one (that is, a *reductio ad absurdum* of the classical type) and it is too long to give here. Moreover, the trigonometric and logarithmic numbers, discussed in the preceding section, also require expression by infinite series if we wish to be able to compute them to any desired degree of precision, and they are likewise transcendental numbers (excluding the exceptions indicated).

There is one number that occurs widely throughout all mathematics which has not been mentioned. It is π, the ratio of the circumference of a circle to its diameter. It is transcendental, although formal proof of that fact was not obtained until efforts had been made for more than two thousand years to prove otherwise — to "square the circle".

In addition to the kinds of transcendental numbers already mentioned, there are many others known to mathematicians. Just as the trigonometric numbers (with the exception already cited) are transcendental, so are other numbers related to other curves, such as the hyperbolic and elliptic functions. So are any mathematical functions involving the number e, since the latter has been shown to be transcendental. Such transcendental numbers include the values of the exponential, beta and gamma functions.

22. Transfinite Numbers. Consider the natural numbers 1, 2, 3, $\cdots$, which are said to be denumerably infinite and compare them with another set, also containing an infinite number of elements. The latter, too, is denumerably infinite if its members can be matched up one-by-one with the natural numbers. Thus, the even numbers form such a set, for the two sets could be matched up as follows: (1, 2), (2, 4), (3, 6), and so on. All sets of this kind are said to have the cardinal number *aleph-null*, written with the Hebrew character, $\aleph_0$ and this is the first transfinite cardinal number.

However, surprising as it seems at first, given a denumerably infinite set of real numbers 1, 2, 3, $\cdots$, there are real numbers in this range, which are not members of the set. These are the transcendental numbers. Since the real numbers in this range have been designated by the transfinite number *aleph-null*, the set including the transcendental numbers, constituting a set of higher order, requires another number. Cantor called the cardinal numbers of this set, C, "*the power of the continuum*," which he believed was equal to $\aleph_1$. Mathematicians today, however, are uncertain about this and there may be other transfinite numbers between $\aleph_0$ and C. It is generally agreed that the following are some properties of these strange numbers (n is a finite number): (1) $\aleph_0 + n = \aleph_0$; (2) $\aleph_0 + \aleph_0 = \aleph_0$; (3) $n\aleph_0 = \aleph_0$; (4) $\aleph_0{}^n = \aleph_0$; (5) $\aleph_0{}^{\aleph_0} = C$; (6) $\aleph_0 + C = C$; (7) $\aleph_0 C = C$. It is also possible to conceive of cardinal numbers $\aleph_0, \aleph_1, \aleph_2, \cdots, \aleph_n$ where n is a natural number and then $\aleph_w, \aleph_{w+1}, \cdots$ larger in turn than any of the previous ones.

23. Groups, Definition of. A group is a set for which a law of combination is defined for members of the set, and such that the products* of this combination are:

1. Every product of two members of the set, or of every member of the set by itself, is a member of the set.

2. The set contains a unit member I for which $Ia = aI = a$, where a is any member of the set.

3. For every member of the set, a, there is an inverse a^{-1}, such that $aa^{-1} = I$ (the unit member).

4. The associative law holds, $a(bc) = (ab)c$.

The set of all integers, positive, negative and zero, form a group if the law of combination is addition. The unit member is 0 (since $0 + a = a + 0 = a$, where a is any integer). The inverse member for every member is its negative (since $a + -a = 0$, where a is any integer). The associative law holds, i.e., $a + (b + c) = (a + b) + c$, where a, b and c are any integers.

Note that the set of all integers, positive, negative and zero do not form a group if the law of combination is multiplication, since in that

---

*The word "product" as used in group theory does not necessarily mean the result of multiplication, but whatever term is obtained by the combination operation defined for the group. Moreover, the notation $a(bc)$, $(ab)c$, etc. does not necessarily mean multiplication, but denotes products obtained by the combination operation of the group.

case, the unit member is 1, because $1 \cdot a = a \cdot 1 = a$, but the member 0 has no inverse, i.e., there is no member x such that $0 \cdot x = 1$. See discussion in preceding section, where multiplication by zero was excluded by definition.

24. Permutation*Groups. An important group is the permutation group, which is a one-to-one function of the group on itself. (See page 00). Thus a three membered permutation group is

$$P = \begin{pmatrix} a & b & c \\ b & c & a \end{pmatrix} \tag{1-48}$$

The top row in this representation of the group contains all the members. The second row gives for each member its function value, that is, b in the second row beneath a in the first row means that a is replaced by b, c in the second row below b in the first means that b is replaced by c, and a in the second row below c in the first means that c is replaced by a. Each of these rows is called a cycle.

The product of two permutation groups (using the word product in the group sense of a combination or in the more precise term, a composition) is found in the following way:

1. Change the order of the members in the first cycle of the second permutation group so that they correspond with the second cycle of the first permutation group.

2. Perform the same transpositions on the second cycle of the second permutation group as those done in (1) on its first cycle. The result is the second cycle of the product permutation group.

3. Write the product permutation group to have as its first cycle the first cycle of the first permutation group, and as its second cycle the cycle found in (2).

Example. Given the two permutation groups

$$P' = \begin{pmatrix} a & b & c \\ a & c & b \end{pmatrix} \quad \text{and} \quad P'' = \begin{pmatrix} a & b & c \\ c & b & a \end{pmatrix} \tag{1-49}$$

To find their product perform numbered steps above as follows:

1. Change order of numbers of the first cycle $(a\ b\ c)$ of the second permutation group (P'') so that they correspond with the second cycle $(a\ c\ b)$ of the first permutation group (P'). This change is $a\ b\ c$.

*Permutations are discussed in algebra texts, such as Thompson, *Algebra for the Practical Man*, 3rd edition, Van Nostrand (1962). They are also reviewed in Chapter 3 in this book, "Probability, Statistics and Quality Control".

2. Perform the same transpositions on the second cycle ($c\ b\ a$) of the second permutation group (P'') as was done in (1) on its first cycle, thus,

$$a\ b\ c \qquad \text{First cycle}$$
$$c\ b\ a \qquad \text{Second cycle}$$

so that $c\ a\ b$ is the second cycle of the product permutation group.

3. Therefore, product permutation group is

$$P''P' = \begin{pmatrix} a & b & c \\ c & a & b \end{pmatrix} \quad \begin{array}{l} \text{(First cycle of first permutation group)} \\ \text{(Cycle found in (2)).} \end{array}$$

Note that permutation products, like matrices, are written with their group symbols in "reverse order," i.e., as $P''P'$. This is done because it is the last term so written (i.e., P') which is transformed by the first (i.e., P'') to yield the product. By extension of this reasoning the triple product of (P' and P'') and P''' would be written $P'''P''P'$.

To develop the subject of permutation groups further, let us write all possible permutation groups having the members $a\ b\ c$. They are:

$$P_A = \begin{pmatrix} a & b & c \\ a & b & c \end{pmatrix} \tag{1-51}$$

(This is the identity permutation, which meets the group requirement for an identity member that combines with any other of these groups to leave it unchanged, i.e., $P_A P_X = P_X P_A = P_X$).

$$P_B = \begin{pmatrix} a & b & c \\ b & c & a \end{pmatrix} \tag{1-52}$$

$$P_C = \begin{pmatrix} a & b & c \\ c & a & b \end{pmatrix} \tag{1-53}$$

(P_C is the product found in above example).

$$P_D = \begin{pmatrix} a & b & c \\ a & c & b \end{pmatrix} \tag{1-54}$$

(This is group P' in above example).

$$P_E = \begin{pmatrix} a & b & c \\ c & b & a \end{pmatrix} \tag{1-55}$$

(This is group P'' in above example).

$$P_F = \begin{pmatrix} a & b & c \\ b & a & c \end{pmatrix} \tag{1-56}$$

Just as we obtained the product group of groups P_D and P_E (then called P' and P'') in the above example, and found it to be P_C, so we can obtain the products of all the other pairs of groups which can be formed from these six permutation groups.   Since six groups or any six distinct objects, yield thirty-six permutation pairs, we will not carry out all their compositions, but simply state that they follow the rule given above, and tabulate the results.   This table, which is called a group table, is constructed by writing the members of the table (which in this case are groups) along the top and left side of a square array, which is then filled in by writing at the intersection of every column and row the product of the group at the top of the column and that at the left end of the row.   Thus, since we found that the product of P_D and P_E was P_C, then P_C is written at the intersection of the column headed P_D and the row labelled P_E as shown below.

	P_A	P_B	P_C	P_D	P_E	P_F
P_A	—	—	—	—	—	—
P_B	—	—	—	—	—	—
P_C	—	—	—	—	—	—
P_D	—	—	—	—	—	—
P_E	—	—	—	—	P_C	—
P_F	—	—	—	—	—	—

(1-57)

The other members of the group table are readily found by the process for composition, which was worked out for P_D and P_E to obtain P_C.   In the case of P_A, the identity group, this is unnecessary, since we know that $P_A P_X = P_X P_A = P_X$, where P_X is any member of the group.   Therefore, for the P_A column and the P_A row of the table, we simply duplicate the symbols as written at the left-hand column and top of the table, respectively.   Following these steps, we obtain the completed table.

	P_A	P_B	P_C	P_D	P_E	P_F
P_A	P_A	P_B	P_C	P_D	P_E	P_F
P_B	P_B	P_C	P_A	P_F	P_D	P_E
P_C	P_C	P_A	P_B	P_E	P_F	P_D
P_D	P_D	P_E	P_F	P_A	P_B	P_C
P_E	P_E	P_F	P_D	P_C	P_A	P_B
P_F	P_F	P_D	P_E	P_B	P_C	P_A

(1-58)

Permutation groups have a number of important applications in science and technology. Two such applications arise in the study of crystal structure and molecular structure. Many of the properties of crystals and molecules are related to their geometric configurations, that is, to the geometrical configurations of the particles (atoms and ions) of which they are composed. A valuable theoretical approach to these configurations is the study of their symmetry properties.

In the dictionary definition, a physical object is said to be symmetric when it is arranged in accordance with a certain similarity with reference to a certain geometrical entity or position, which may be a point, line or plane. In mathematics, operational definitions of symmetry are used. Thus, a center of symmetry is a point such that any line drawn through it intersects a surface of the object (or the bounding surface of the set points) at equal distances on either side. An axis of symmetry is a line drawn within a body or set of points in such a position and direction that a rotation of the body through an angle $2\pi/n$ radians about the line as an axis, n being an integer, results in a configuration indistinguishable from the original. A plane of symmetry is a plane passed through a body or set of points in such a position and direction that the reflection of the body or set of points in the plane results in a configuration indistinguishable from the original.

Now let us investigate the symmetry of the equilateral triangle. In Figure 1-11, let us, rotate the equilateral triangle abc counterclockwise about the Z-axis, which is perpendicular to the plane of the paper at 0,

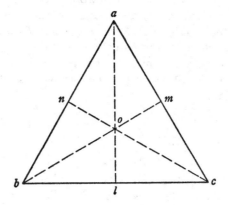

Fig. 1-11.

the intersection of the three bisectors al, bm and cn.   When the rotation
has been continued until point a has moved to the position originally
occupied by b, then point b has moved over the former position of c,
and point c has moved over the former position of a.   (Angular amount
of rotation $= 120°$).   This rotation may be then represented by the
permutation group

$$\begin{pmatrix} a & b & c \\ b & c & a \end{pmatrix} \tag{1-59}$$

which was P_B in the list developed on Page 28.   Continuing the rotation to the next position of coincidence, we have a over former position
of c, b over the former position of a, and c over the former position of b,
the group being

$$\begin{pmatrix} a & b & c \\ c & a & b \end{pmatrix} \tag{1-60}$$

which was the former P_C.
Now continuing the rotation one step further, we find the next point
of coincidence to restore the triangle to its original position.   In other
words, a complete rotation (through $360°$) gives the identity permutation group

$$P_A = \begin{pmatrix} a & b & c \\ a & b & c \end{pmatrix} \tag{1-61}$$

Since the equilateral triangle had these (three) symmetry operations
in its rotation about the Z-axis, the latter is called the axis of symmetry.
Suppose now that we turn the triangle over.   This symmetry operation
is called reflection (or inversion) and the axis about which the turning
is done is called the axis of reflection.   If the triangle is inverted about
axis al, then point a remains unchanged, and b and c change places, so
that the permutation group is

$$\begin{pmatrix} a & b & c \\ a & c & b \end{pmatrix} \tag{1-62}$$

which is the group P_D of Page 28.   Similarly, reflection (inversion
about) axis bm yields

$$P_E = \begin{pmatrix} a & b & c \\ c & b & a \end{pmatrix} \tag{1-63}$$

and reflection in axis cn yields

$$P_F = \begin{pmatrix} a & b & c \\ b & a & c \end{pmatrix} \tag{1-64}$$

Now let us consider the result of performing two symmetry operations successively, for example, reflection in axis bm, rotation through 120°. Then as above, reflection in axis bm yields the position shown in Figure 1-12. Then rotation through 120° moves a to the position occupied

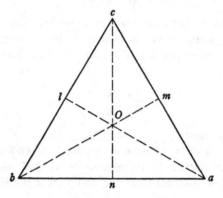

Fig. 1-12.

by c in Figure 1-12, c to the position occupied by b in Figure 1-12, and b to the position occupied by a in Figure 1-12, the new arrangement being shown in Figure 1-13. But we have shown that the reflection

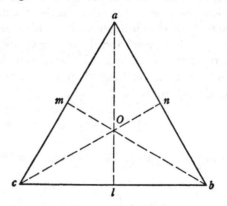

Fig. 1-13.

performed to give the position shown in Figure 1-12 may be represented by the group P_E, and that the rotation performed to give the position of Figure 1-13 from that in Figure 1-12 may be represented by P_B. We also see from Figure 1-13 and Page 28 that the position is represented by P_D. But from the table on Page 29, the composition of P_E and P_B gives P_D. Therefore, the permutation groups correspond to symmetry operations, and the result of successive operations may be found from the product of the corresponding permutation groups.

Of course, the symmetry operations and corresponding groups products that have been described up to this point apply only to the equilateral triangle. The rectangle would have only four symmetry operations, shown in Figure 1-14. Two are rotations, the second being represented by the identity group, and two are reflections. Since there are only two symmetry operations possible in rotation about the Z-axis, the latter is called a twofold axis of symmetry.

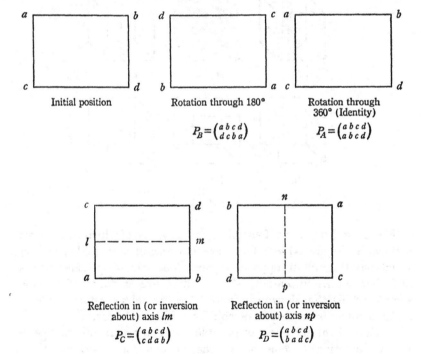

Initial position

Rotation through 180°

$$P_B = \begin{pmatrix} a\,b\,c\,d \\ d\,c\,b\,a \end{pmatrix}$$

Rotation through 360° (Identity)

$$P_A = \begin{pmatrix} a\,b\,c\,d \\ a\,b\,c\,d \end{pmatrix}$$

Reflection in (or inversion about) axis *lm*

$$P_C = \begin{pmatrix} a\,b\,c\,d \\ c\,d\,a\,b \end{pmatrix}$$

Reflection in (or inversion about) axis *np*

$$P_D = \begin{pmatrix} a\,b\,c\,d \\ b\,a\,d\,c \end{pmatrix}$$

Fig. 1-14.

Since crystals, molecules and other physical objects are three-dimensional, the number of symmetry elements in them may be large, especially for the regular figures (elements is the accepted name for symmetry operations; we have used the latter term up to this point to avoid confusion with permutation group members which are also called elements). Thus the cube may be shown to have twenty-three such elements. There are three rectangular planes of symmetry passing through the mid-points of the sides — Figure 1-15a shows the YZ plane of symmetry, and the XY and XZ planes are also symmetry elements. There are six diagonal planes of symmetry, like that shown in Figure 1-15b, of which there is one such plane mounted on the diagonals of each pair of opposite faces (three pairs of opposite faces, two diagonals each).

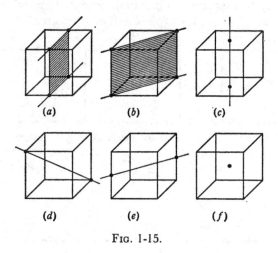

(a) (b) (c)

(d) (e) (f)

Fig. 1-15.

There are three axes of four-fold symmetry, one of which (the Z-axis) is shown in Figure 1-15c. There are four axes of three-fold symmetry, which pass through the opposite vertices (four pairs of opposite vertices) like that shown in Figure 1-15d. There are six axes of two-fold symmetry, like that shown in Figure 1-15e (six pairs of opposite edges); and last, there is one center of symmetry (Figure 1-15f).

In view of the members of possible symmetry elements in crystals, various notation systems have been devised for describing them. That of Hermann-Manguin uses arabic numbers to designate rotation about axes, the magnitude of the number denoting the multiplicity of the axis, (six for six-fold, four for four-fold, etc.).   In this system, an inversion axis is considered to combine rotation about an axis with

TABLE OF SYMMETRY POINT GROUPS
Herrnmann Manguin Symbols

System	Crystallographic Elements	Point Groups
Cubic or regular	Three axes at right angles: all equal $\alpha = \beta = \gamma = 90°$ $a : b : c = 1 : 1 : 1$	23 $m3$ 432 $\overline{4}3m$ $m3m$
Tetragonal	Three axes at right angles: two equal $\alpha = \beta = \gamma = 90°$ $a : b : c = 1 : 1 : y$	4 $\overline{4}$ $4/m$ 422 $4mm$ $\overline{4}2m$ $4/mmm$
Orthorhombic or Rhombic	Three axes at right angles: unequal $\alpha = \beta = \gamma = 90°$ $a : b : c = x : 1 : y$	222 $mm2$ mmm
Monoclinic	Three axes: one pair not at right angles: unequal $\alpha = \gamma = 90°$ $\beta \neq 90°$ $a : b : c = x : 1 : y$	2 m $2/m$
Triclinic or Anorthic	Three axes not at right angles $\alpha, \beta, \gamma \neq 90°$ $a : b : c = x : 1 : y$	1 $\overline{1}$
Hexagonal	Three axes coplanar at 60°: equal Fourth axis at right angles to other three $a_1 : a_2 : a_3 : b = 1 : 1 : 1 : x$	6 $\overline{6}$ $6/m$ 622 $6mm$ $\overline{6}m2$ $6/mmm$
Trigonal or Rhombohedral	Three axes equally inclined: not at right angles: all equal $\alpha = \beta = \gamma \neq 90°$ $a : b : c = 1 : 1 : 1$	3 $\overline{3}$ 32 $3m$ $\overline{3}m$

inversion with respect to a point on it.   Inversion axes are designated by a bar over the multiplicity number, $\bar{6}$ for a six-fold inversion, $\bar{4}$ for four-fold, etc.).   However, the symbol $\bar{2}$ does not appear, for the reason that $\bar{2}$ is a reflection plane and is denoted by m.   A mirror plane perpendicular to an axis is written after it following a diagonal mark, as e.g., $4/m$ if the axis is a four-fold rotation axis.   The order $m4$ would mean that the four-fold axis was neither perpendicular to the plane or in it.   Then $4/mm$ would mean a four-fold axis, four planes passing through it and one perpendicular to it.

In this system, point groups of the seven crystallographic systems are designated in the table on Page 35.

In this table, the crystallographic elements have been included for completeness, even though they are not part of the subject matter of this chapter.   They are the angles between the crystallographic axes, and the ratios of the intercepts upon them cut by the standard plane of the crystal.

While the foregoing thirty-two point groups are basic to the study of the geometry of crystals, and the properties which depend upon the geometry, it must be considered that the point group is merely a unit in the overall space lattice.   The latter results from the repetition of the point-group pattern in the three directions, and the operation of extending it requires the addition of two other symmetry elements, glide planes and screw axes.   By the various applications of these two elements to the 32 point groups, there is produced a total of 230 space groups, which take into consideration the internal structure of the crystal, as well as its external form.   Notations have been developed for the space groups, just as they have for the point groups, and are explained in books on crystallography.

Problems for Solution

1. For every set S, what are $S \cap E$, $S \cup E$, $S \cap S$ and $S \cup S$ where E is the universal set?

2. For every set S, what are $S' \cap E$, $S' \cup E$, $S' \cap S$ and $S' \cup S$, where S' is the complement of S and E is the universal set?

3. For every set S, what are $S \cup \phi$ and $S' \cup \phi$, where ϕ is the empty set and S' is the complement of S?

4. Draw the diagrams for $A \cup B$, $A \cap B$ and $A \supset B$.

5. Write the complete list of the subsets of a, b, c, d, e.

6. Express the position of a point which has the coordinates 5, 7, 12 and 4 in four-dimensional space.

7. Express the reflexive relation of every member d of D.
8. Express the symmetric relation between d and e.
9. Express the transitive relation between d, e and f.
10. Under what operations is the set $-1, 0, 1$ closed?
11. Rationalize the decimal numbers

$$.9\overline{09}; \; .0073\overline{185}; \; .\overline{142857}$$

12. Which of the following operations yield rational numbers:

   (a) $\sqrt{3} + \sqrt{3}$ (d) $\sqrt{8} \div 2$

   (b) $\sqrt{3} \times \sqrt{3}$ (e) $\sqrt{8} \div \sqrt{2}$

   (c) $\sqrt{3} \div \sqrt{3}$

13. Which of the following trigonometric functions are rational?

   (a) cos 60°   (d) cosec 30°

   (b) cotan 45°   (e) sec 30°

   (c) sin 60°   (f) cos 120°

14. Which of the following numbers have rational logs to base 10:

   (a) 3   (d) 10

   (b) 4   (e) 100,000

   (c) 5   (f) .0001

15. What are the four requirements necessary for set to be a group?
16. What is a permutation group?
17. What are the product permutation groups of the following?

$$\begin{pmatrix} a & b & c \\ c & b & a \end{pmatrix}\begin{pmatrix} a & b & c \\ b & a & c \end{pmatrix};$$

$$\begin{pmatrix} a & b & c \\ b & c & a \end{pmatrix}\begin{pmatrix} a & b & c \\ c & a & b \end{pmatrix};$$

$$\begin{pmatrix} a & b & c \\ a & c & b \end{pmatrix}\begin{pmatrix} a & b & c \\ b & a & c \end{pmatrix}$$

18. What are the product permutation groups of the following?

$$\begin{pmatrix} a & b & c & d \\ a & d & b & c \end{pmatrix}\begin{pmatrix} a & b & c & d \\ a & c & b & d \end{pmatrix};$$

$$\begin{pmatrix} a & b & c & d \\ d & c & b & a \end{pmatrix}\begin{pmatrix} a & b & c & d \\ b & a & d & c \end{pmatrix};$$

$$\begin{pmatrix} a & b & c & d \\ b & c & d & a \end{pmatrix}\begin{pmatrix} a & b & c & d \\ b & d & a & c \end{pmatrix}$$

19. Write the symmetry groups of the regular hexagon.

Chapter 2
MATRICES AND DETERMINANTS

In ordinary arithmetic, we are concerned primarily with single numbers, which are subject to various operations, i.e. addition, multiplication, subtraction, and division, to produce new numbers. In matrix algebra, this process is extended to collections of numbers. One type of collection of numbers is the simple sequence, which may be a pair of real numbers, e.g. 7, 5 or a_1, a_2; a trio of real numbers, e.g. 7, 5, 9 or a_1, a_2, a_3, or any larger sequence, e.g., a_1, a_2, $a_3 \cdots a_n$ of real numbers,

The pair of real numbers is exemplified by a two-dimensional vector in which the two numbers may function to define the magnitude and direction of a velocity, force or other directed quantity, or of their geometric representation. Similarly, a trio of real numbers is exemplified by a three-dimensional vector.* In fact, the concept of a vector may be generalized to mean a sequence of numbers, which may consist of two, three or more members.

Furthermore, just as mathematicians extend their interest from two and three dimensional vectors to n-dimensional ones, so they proceed from vectors (i.e. numbers in a single row or column) to those in more than one row or column. Such collections of numbers may have a number of forms: however, the matrices which are the subject to this chapter are either rectangular, as

$$A = \begin{pmatrix} A_{11} & A_{12} & A_{13} & A_{14} & A_{15} \\ A_{21} & A_{22} & A_{23} & A_{24} & A_{25} \\ A_{31} & A_{32} & A_{33} & A_{34} & A_{35} \end{pmatrix} \qquad (2\text{-}1)$$

or square, as

$$B = \begin{pmatrix} B_{11} & B_{12} & B_{13} & B_{14} \\ B_{21} & B_{22} & B_{23} & B_{24} \\ B_{31} & B_{32} & B_{33} & B_{34} \\ B_{41} & B_{42} & B_{43} & B_{44} \end{pmatrix} \qquad (2\text{-}2)$$

*For a discussion of two- and three-dimensional vectors, see *Algebra for the Practical Man*, by Thompson.

They may have any number of rows and columns, provided that the rectangular matrix has the same number of rows throughout, and the same number of columns throughout, and provided that the square matrix has the same number of both rows and columns throughout.

The notation used above, that is, the use of parentheses to enclose the members of a matrix, the use of an italic capital letter for a matrix, and the use of that same letter followed by two subscript numbers for a member (called an element) of a matrix is widely followed.* The first subscript number designates the row to which the element belongs, and the second number its column. Thus, A_{43} designates the element in the fourth row and third column of matrix A. This general notation is used in stating the rules for the various operations on matrices, which is accompanied in every case by examples worked with matrices having real numbers as their elements.

These rules are an essential part of the concept of a matrix, which without them is a meaningless array. With them, however, it becomes a mathematical entity, which has a wide range of applications. The first of these rules to be discussed is that for addition of matrices.

25. Addition of Matrices. The operation of adding two matrices consists of adding each element of one matrix to the corresponding element of the other. In other words, in order to add two matrices it is required that they have the same number of elements, arranged in the same number of rows and columns. Thus we say that the addition of two matrices requires that there be M rows in each, N columns in each, and therefore $M \times N$ elements in each. If the matrix is a square one, then $M = N$.

Example 1.   Add the matrices A and B, which follow:

$$A = \begin{pmatrix} 1 & 2 & 3 \\ 4 & 5 & 6 \end{pmatrix} \qquad B = \begin{pmatrix} 7 & 8 & 9 \\ 10 & 11 & 12 \end{pmatrix}$$

Then we have

$$A + B = \begin{pmatrix} 1 & 2 & 3 \\ 4 & 5 & 6 \end{pmatrix} + \begin{pmatrix} 7 & 8 & 9 \\ 10 & 11 & 12 \end{pmatrix}$$

$$= \begin{pmatrix} 1+7 & 2+8 & 3+9 \\ 4+10 & 5+11 & 6+12 \end{pmatrix}$$

$$= \begin{pmatrix} 8 & 10 & 12 \\ 14 & 16 & 18 \end{pmatrix}$$

*The use of bold-face letters for matrices is also found in many publications.

Or $A + B = C$, where C is the above matrix.

This addition operation can be represented symbolically as

$$C = A + B = \begin{pmatrix} A_{11} & A_{12} & A_{13} \\ A_{21} & A_{22} & A_{23} \end{pmatrix} + \begin{pmatrix} B_{11} & B_{12} & B_{13} \\ B_{21} & B_{22} & B_{23} \end{pmatrix}$$

$$= \begin{pmatrix} A_{11} + B_{11} & A_{12} + B_{12} & A_{13} + B_{13} \\ A_{21} + B_{21} & A_{22} + B_{22} & A_{23} + B_{23} \end{pmatrix}$$

(2-3)

The procedure followed for the addition of matrices is extended to the operation of subtraction by employing the following method. To subtract matrix B from matrix A, it is first necessary to multiply each element in the former by (-1). Then the operation of matrix addition is applied to the two matrices in order to obtain their difference. Thus the form used for the subtraction of matrices is $A + (-B)$, which is the algebraic equivalent of the expression $A - B$.

The following example provides an illustration of this operation.

Example 2. Subtract matrix B from matrix A.

$$A = \begin{pmatrix} 10 & 6 & 7 \\ 8 & 5 & 3 \end{pmatrix} \qquad B = \begin{pmatrix} 9 & 4 & 2 \\ 6 & 3 & 1 \end{pmatrix}$$

then we have

$$A - B = \begin{pmatrix} 10 & 6 & 7 \\ 8 & 5 & 3 \end{pmatrix} - \begin{pmatrix} 9 & 4 & 2 \\ 6 & 3 & 1 \end{pmatrix}$$

$$= \begin{pmatrix} 10 & 6 & 7 \\ 8 & 5 & 3 \end{pmatrix} + \begin{pmatrix} -9 & -4 & -2 \\ -6 & -3 & -1 \end{pmatrix}$$

$$= \begin{pmatrix} 1 & 2 & 5 \\ 2 & 2 & 2 \end{pmatrix}$$

Or $A + (-B) = C$.   (2-4)

It should be remembered that the rules stated above, governing the form of matrices which can be added, apply also to determining whether the subtraction operation can be employed. Matrices of dissimilar dimensions can be neither added nor subtracted. For example, neither operation may be used to combine the two matrices:

$$\begin{pmatrix} 5 & 8 \\ 1 & 2 \end{pmatrix} \qquad \begin{pmatrix} 1 & 9 & 6 & 5 \\ 3 & 4 & 3 & 7 \end{pmatrix}$$

26. Scalar Multiplication of Matrices. The operation of multiplying a given matrix by a scalar consists of multiplying every element in the matrix by the scalar quantity. In contrast to the rules specified for the above operations of addition and subtraction of matrices, there are no restrictions on the form of a matrix that may undergo the process of multiplication by a scalar.

Example 1. Multiply the matrix A by the scalar $r = 2$.

$$A = \begin{pmatrix} 3 & 4 & 6 & 2 \\ 9 & 2 & 7 & 5 \end{pmatrix} \qquad r = 2$$

Then $rA = 2 \begin{pmatrix} 3 & 4 & 6 & 2 \\ 9 & 2 & 7 & 5 \end{pmatrix}$

$$= \begin{pmatrix} 6 & 8 & 12 & 4 \\ 18 & 4 & 14 & 10 \end{pmatrix}$$

Or $rA = C$. It can be seen from this example that the product of a matrix and a scalar is always another matrix.

Written symbolically this operation may be reduced to the following formula.

$$r \times \begin{pmatrix} A_{11} & A_{12} & A_{13} & A_{14} \\ A_{21} & A_{22} & A_{23} & A_{24} \\ A_{31} & A_{32} & A_{33} & A_{34} \end{pmatrix} = \begin{pmatrix} rA_{11} & rA_{12} & rA_{13} & rA_{14} \\ rA_{21} & rA_{22} & rA_{23} & rA_{24} \\ rA_{31} & rA_{32} & rA_{33} & rA_{34} \end{pmatrix} \qquad (2\text{-}5)$$

Certain of the laws that apply to addition in arithmetic and elementary algebra apply also to the corresponding matrix operations.

I. The associative law which is expressed in algebraic terms as

$$(a + b) + c = a + (b + c)$$

is also applicable to matrices in the form

$$(A + B) + C = A + (B + C) \qquad (2\text{-}6)$$

where A, B, and C are matrices having the same number of rows and the same number of columns.

II. Similarly, the commutative law of addition which can be stated in algebraic terms as:

$$a + b = b + a$$

is applicable to matrices according to the formula

$$A + B = B + A \qquad (2\text{-}7)$$

where A and B are again matrices, having the same number of rows and same number of columns.

III. The distributive law, which is stated algebraically as

$$c(a + b) = ac + bc$$

also supplies to multiplication of matrices by a scalar;

$$r(A + B) = rA + rB \tag{2-8}$$

for use in operations with matrices. Here A and B are matrices and r is a scalar quantity.

27. Multiplication of Matrices by Matrices. Matrix multiplication is defined as *row by column* multiplication and is conducted in a systematic and very definite way. Two matrices A, B may be multiplied together to form the product AB only when the number of columns in A is equal to the number of rows in B. In the product matrix AB the first element of the first row is obtained by multiplying the elements of the first row of A by the corresponding elements of the first column of B and summing the products. Similarly, the second element in the first row of the product matrix is obtained by multiplying the elements of the first row of A by the corresponding elements of the second column of B and again summing the results. Likewise to obtain the first element in the second row of the product matrix AB, we must multiply the elements of the second row of A by the corresponding elements of the first column of B, and add the products. The remaining elements in the product matrix are found by continuing this process.

Before we proceed any further, let us consider some examples that should help to clarify this operation.

Example 1.   Multiply matrix A by matrix B

$$A = \begin{pmatrix} 9 & 5 \\ 4 & 2 \end{pmatrix} \qquad B = \begin{pmatrix} 6 & 7 & 8 \\ 3 & 10 & 12 \end{pmatrix}$$

To obtain the first element in the product matrix AB we are concerned only with the elements of A and B that are represented below:

$$\begin{pmatrix} 9 & 5 \\ \cdot & \cdot \end{pmatrix} \qquad \begin{pmatrix} 6 & \cdot & \cdot \\ 3 & \cdot & \cdot \end{pmatrix}$$

According to the preceding explanation the first element of AB is found by multiplying 9 by 6 and by multiplying 5 by 3 and then

summing the two products. Thus we have $54 + 15 = 69$ and this
sum becomes the first element in the first row of AB as shown below.

$$\begin{pmatrix} 69 & \cdot & \cdot \\ \cdot & \cdot & \cdot \end{pmatrix}$$

To obtain the second element in the first row of AB we are concerned
only with the following elements of A and B

$$\begin{pmatrix} 9 & 5 \\ \cdot & \cdot \end{pmatrix} \begin{pmatrix} \cdot & 7 & \cdot \\ \cdot & 10 & \cdot \end{pmatrix}$$

By following the procedure used in obtaining the first element in
AB we have

$$(9 \times 7) + (5 \times 10) = 63 + 50 = 113$$

Thus the partial product matrix AB now appears as

$$\begin{pmatrix} 69 & 113 & \cdot \\ \cdot & \cdot & \cdot \end{pmatrix}$$

To complete the first row of the product matrix we perform the same
operation on the first row of A and the third column of B. Thus we
have:

$$(9 \times 8) + (5 \times 12) = 72 + 60 = 132$$

The first row of AB is then seen to be

$$\begin{pmatrix} 69 & 113 & 132 \\ \cdot & \cdot & \cdot \end{pmatrix}$$

To obtain the first element in the second row of AB, we need employ
only the elements of A and B shown below

$$\begin{pmatrix} \cdot & \cdot \\ 4 & 2 \end{pmatrix} \begin{pmatrix} 6 & \cdot & \cdot \\ 3 & \cdot & \cdot \end{pmatrix}$$

Again by following the standard procedure we have:

$$(4 \times 6) + (2 \times 3) = 24 + 6 = 30$$

When this result is inserted into AB, the partial product matrix
appears as

$$\begin{pmatrix} 69 & 113 & 132 \\ 30 & \cdot & \cdot \end{pmatrix}$$

It is evident that the remaining elements of AB can be found by following the same procedure of row by column multiplication. Thus AB becomes

$$\begin{pmatrix} 69 & 113 & 132 \\ 30 & [(4 \times 7) + (2 \times 10)] & [(4 \times 8) + (2 \times 12)] \end{pmatrix}$$

$$= \begin{pmatrix} 69 & 113 & 132 \\ 30 & 48 & 56 \end{pmatrix} = AB$$

Example 2a. Multiply matrix A by matrix B, as given below

$$A = \begin{pmatrix} 2 & 4 & 6 \\ 8 & 10 & 12 \\ 14 & 16 & 18 \end{pmatrix} \qquad B = \begin{pmatrix} 1 \\ 3 \\ 5 \end{pmatrix}$$

$$AB = \begin{pmatrix} (2 \times 1) + (4 \times 3) + (6 \times 5) \\ (8 \times 1) + (10 \times 3) + (12 \times 5) \\ (14 \times 1) + (16 \times 3) + (18 \times 5) \end{pmatrix}$$

$$= \begin{pmatrix} 2 + 12 + 30 \\ 8 + 30 + 60 \\ 14 + 48 + 90 \end{pmatrix}$$

$$= \begin{pmatrix} 44 \\ 98 \\ 152 \end{pmatrix}$$

Example 2b. Using the same matrices A and B as those given in the preceding example, attempt to multiply B by A to arrive at the matrix product BA.

$$\begin{pmatrix} 1 \\ 3 \\ 5 \end{pmatrix} \times \begin{pmatrix} 2 & 4 & 6 \\ 8 & 10 & 12 \\ 14 & 16 & 18 \end{pmatrix} = ?$$

We now see that it is impossible to conduct such an operation for *the number of columns in B is not equal to the number of rows in A.* This example illustrates an important difference between the rules of ordinary algebraic multiplication and those of matrix multiplication. In arithmetic or elementary algebra multiplication is commutative, that is

$$ab = ba$$

But in the multiplication of two matrices in most cases

$$AB \neq BA$$

and in many cases AB has no complementary product BA, as in Example 2b above. The following example illustrates one case in which both AB and BA exist.

Example 3.   Multiply matrix A below by matrix B below to find AB and then multiply matrix B by matrix A to find BA.

NOTE.   Matrix A and matrix B are here square matrices having the same dimensions.

$$A = \begin{pmatrix} 1 & 2 \\ 3 & 4 \end{pmatrix} \qquad B = \begin{pmatrix} 5 & 6 \\ 7 & 8 \end{pmatrix}$$

$$AB = \begin{pmatrix} [(1 \times 5) + (2 \times 7)][(1 \times 6) + (2 \times 8)] \\ [(3 \times 5) + (4 \times 7)][(3 \times 6) + (4 \times 8)] \end{pmatrix}$$

$$AB = \begin{pmatrix} 19 & 22 \\ 43 & 50 \end{pmatrix}$$

$$BA = \begin{pmatrix} [(5 \times 1) + (6 \times 3)][(5 \times 2) + (6 \times 4)] \\ [(7 \times 1) + (8 \times 3)][(7 \times 2) + (8 \times 4)] \end{pmatrix}$$

$$BA = \begin{pmatrix} 23 & 34 \\ 31 & 46 \end{pmatrix}$$

$$AB \neq BA$$

It may be noted that there are special cases in which, for two matrices A and B, both AB and BA do exist and are equal.

The following symbolic expressions correspond to the three examples that have been worked out above. The first equation illustrates the procedure used in the multiplication of a 2×2 matrix by a 2×3 matrix. This operation was performed in example 1 using real numbers. Likewise, equation 2 corresponds to the second example used above and represents the general procedure for multiplying a 3×3 matrix by a 3×1 matrix. The third equation corresponds to the third example and shows symbolically the multiplication of two square matrices.

Example 1. $\begin{pmatrix} a & b \\ c & d \end{pmatrix} \begin{pmatrix} e & f & g \\ h & i & j \end{pmatrix} = \begin{pmatrix} (ae + bh) & (af + bi) & (ag + bj) \\ (ce + dh) & (cf + di) & (cg + dj) \end{pmatrix}$

$$(2\text{-}9)$$

Example 2a.
$$\begin{pmatrix} a & b & c \\ d & e & f \\ g & h & i \end{pmatrix} \begin{pmatrix} j \\ k \\ l \end{pmatrix} = \begin{pmatrix} aj + bk + cl \\ dj + ek + fl \\ gj + hk + il \end{pmatrix} \tag{2-10}$$

Example 3.
$$\begin{pmatrix} a & b \\ c & d \end{pmatrix} \begin{pmatrix} e & f \\ g & h \end{pmatrix} = \begin{pmatrix} (ae + bg) & (af + bh) \\ (ce + dg) & (cf + dh) \end{pmatrix} \tag{2-11}$$

We have now seen that any matrix A may be multiplied by a second matrix B to form the product AB if the matrices A and B have the following configurations.

$$A = \begin{pmatrix} A_{11} & A_{12} & \cdots & A_{1n} \\ A_{21} & A_{22} & \cdots & A_{2n} \\ \cdot & \cdot & \cdots & \cdot \\ \cdot & \cdot & \cdots & \cdot \\ A_{m1} & A_{m2} & \cdots & A_{mn} \end{pmatrix} \quad B = \begin{pmatrix} B_{11} & B_{12} & \cdots & B_{1p} \\ B_{21} & B_{22} & \cdots & B_{2p} \\ \cdot & \cdot & \cdots & \cdot \\ \cdot & \cdot & \cdots & \cdot \\ \cdot & \cdot & \cdots & \cdot \\ B_{n1} & B_{n2} & \cdots & B_{np} \end{pmatrix}$$

$$\tag{2-12}$$

NOTE. The one condition that must be met is that the number of columns in the first matrix be equal to the number of rows in the second as indicated above by the second subscript n in the elements of the last column of the first matrix A being identical with the first subscript n in the last row of the second matrix B.

28. Application of Matrices. Matrices have long been used in scientific and engineering problems, such as those arising from the formulation of atomic and crystal structure, or any others involving complex many-part systems. More recently, matrix problems have come into use in industrial production calculations, especially those relating to the maximization of profit. This development is due primarily to the greater versatility of certain manufacturing plants, and hence to the greater range of choice presented in their production planning.

A case in point is the modern oil refinery. Once the preliminary operations have been performed on the crude petroleum, the intermediate products can be processed and combined in a number of ways to yield a variety of products. While the production over a period of time for an entire refining company may be determined largely by market conditions, there still remains a considerable range of choice of products for a given refinery at a given time. Since different products require different quantities of the basic intermediate products, which have different unit costs, and since the final products command different prices, the computation of profit is a matrix calculation. Since the determination of the maximum possible profit requires specific knowledge of limiting (boundary) conditions, and many matrix calculations, it cannot be given here.

29. Determinants. The word *determinant* has frequently been used interchangeably with the word *matrix* in elementary textbooks and this oversight has led to a certain amount of unnecessary confusion. In the preceding sections of this chapter we have seen that a matrix is an array of numbers or other mathematical objects that are arranged in rows and columns in a rectangular (or square) form.

$$\begin{pmatrix} A_{11} & \cdots & A_{1n} \\ \cdot & \cdots & \cdot \\ \cdot & \cdots & \cdot \\ \cdot & \cdots & \cdot \\ A_{m1} & \cdots & A_{mn} \end{pmatrix}$$

The *determinant* or *the determinant of a square matrix* is written by placing the expression det before the matrix, or enclosing its elements in straight lines instead of curved lines. This determinant notation means that the elements of the matrix are to be combined in a defined manner, which is explained below. Matrices which are not square do not have determinants.

The determinant of an $n \times n$ matrix is defined as the sum of $n!$ terms where each term is the product of the elements in a unique combination formed by selecting n elements from the given matrix, none of which stand in the same row or column. Each term that is to be added in finding the determinant of a matrix is preceded by the signs $+$ or $-$ depending on whether the *permutation* of the column subscripts of the chosen elements is even or odd. This sign will be $+$ if the column subscripts form an even permutation of the integers 1, 2, 3 $\cdots$; it will be $-$ if the column subscripts form an odd permutation of the integers 1, 2, 3 $\cdots$. This rule will now be illustrated.

When a set of numbers 1, 2, 3, 4 is arranged as it is here in ascending numerical order, we say that no inversions are necessary to arrange the elements in normal order. However, if these numbers are arranged in any other order, such as 4 2 1 3, we now say that we have a permutation of the elements of the original set. This permutation is referred to as "even" or "odd" depending upon whether the smallest number of inversions needed to return the elements to normal order is even or odd. An *inversion* is simply the reversal in order of two adjacent integers, and the minimum number of inversions needed to bring the elements of a permutation into normal order is called the *index* of that permutation.

Example: Find the index of the following permutation.

$$4 \quad 2 \quad 1 \quad 3$$

First inversion   ②④ 1   3

Second inversion   2 ①④ 3

Third inversion   2   1 ③④

Fourth inversion ①② 3   4

Thus the index of this permutation is 4 and the permutation is said to be even, for 4 is an even integer.

In the following illustration we see how this operation is applied in the calculation of the determinant of a 3 × 3 matrix.   We find it more convenient to denote the order of the rows by the letters a, b, c and the order of the columns by the integers 1, 2, 3.

Example: Find the determinant of matrix A

$$A = \begin{pmatrix} a_1 & a_2 & a_3 \\ b_1 & b_2 & b_3 \\ c_1 & c_2 & c_3 \end{pmatrix} \tag{2-15}$$

In this case $n = 3$, since the given matrix is a 3 × 3, and thus we may expect to find 3! ($= 1 \cdot 2 \cdot 3$) or 6 terms which must be added to give the desired determinant.   In each case we select an "a" element first, a "b" element second and a "c" element third, for **the significance of the permutation of the column subscripts depends upon a uniform order of selection,** and we choose the elements so that no two elements of the same term are taken from the same row or the same column. Hence, we find that the six terms are:

$$(a_1 b_2 c_3)$$
$$(a_2 b_3 c_1)$$
$$(a_3 b_1 c_2)$$
$$(a_1 b_3 c_2) \tag{2-16}$$
$$(a_2 b_1 c_3)$$
$$(a_3 b_2 c_1)$$

To determine the sign of each of the above terms we count the number of inversions that must be made in each permutation of the column subscripts to arrange them in normal order.

TERM	PERMUTATION			INVERSIONS	SIGN	
$(a_1b_2c_3)$	1	2	3	0	+ (note*)	
$(a_2b_3c_1)$	2	3	1	2	+	
$(a_3b_1c_2)$	3	1	2	2	+	(2-17)
$(a_1b_3c_2)$	1	3	2	1	−	
$(a_2b_1c_3)$	2	1	3	1	−	
$(a_3b_2c_1)$	3	2	1	3	−	

Therefore, the determinant of matrix A is $(a_1b_2c_3) + (a_2b_3c_1) + (a_3b_1c_2) - (a_1b_3c_2) - (a_2b_1c_3) - (a_3b_2c_1)$.

30. Finding the Determinants of 2 $\times$ 2 and 3 $\times$ 3 Matrices. To facilitate the process of finding the determinants of 2 $\times$ 2 and 3 $\times$ 3 matrices, mathematicians have devised a number of useful shortcuts. In the following discussion we refer to the diagonal of a square matrix which begins in the upper left hand corner and terminates in the bottom right hand corner as the *main diagonal*. The diagonal beginning in the upper right hand corner and terminating in the lower left hand corner will be referred to as the *secondary diagonal*.

$$\begin{pmatrix} a & b & c \\ d & e & f \\ g & h & i \end{pmatrix} \begin{matrix} \text{----▶ Secondary diagonal} \\ \\ \text{----▶ Main diagonal} \end{matrix} \qquad (2\text{-}18)$$

The determinant of a 2 $\times$ 2 matrix is found by taking the product of the elements lying along the secondary diagonal and subtracting this quantity from the product obtained by multiplying the elements lying along the main diagonal. This operation is illustrated below:

$$\det \begin{pmatrix} 3 & 4 \\ 1 & 2 \end{pmatrix} = (3 \times 2) - (4 \times 1) = (6 - 4) = 2$$

Obviously, this shortcut yields the same result as the general rule, which for a 2 $\times$ 2 determinant would be

$$\det \begin{pmatrix} a_1 & b_1 \\ a_2 & b_2 \end{pmatrix} = a_1b_2 - a_2b_1 \qquad (2\text{-}19)$$

*In counting inversions to determine the sign of a product term, 0 is regarded as an even number.

Evaluating the determinant of a 3 × 3 matrix involves, as noted above, six products. The most convenient method for performing this operation consists of rewriting the determinant followed by a repetition of the first two columns. Thus to find:

$$\det \begin{pmatrix} 1 & 2 & 3 \\ 4 & 5 & 6 \\ 7 & 8 & 9 \end{pmatrix}$$

Write

$$\det \begin{pmatrix} 1 & 2 & 3 \\ 4 & 5 & 6 \\ 7 & 8 & 9 \end{pmatrix} \begin{matrix} 1 & 2 \\ 4 & 5 \\ 7 & 8 \end{matrix}$$

Then draw diagonals as follows:

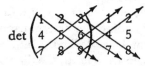

The products of the elements along each of these diagonals are then found, and the sum of those with arrowheads at the top (i.e. along the secondary diagonal and lines parallel to it) is subtracted from the sum of those with arrowheads at the bottom (i.e. along the main diagonal and lines parallel to it).

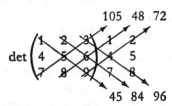

$$(45 + 84 + 96) - (105 + 48 + 72) = 225 - 225 = 0$$

The value of this determinant is 0. This value should be checked by applying the general formula for evaluating 3 × 3 determinants, as given on Page 48.

31. Properties of Matrices and Determinants. Matrices have a number of properties that are useful in the evaluation of their determinants. The most significant of these are explained below:

First Property — If all the elements in a row or a column of a square matrix are zero, the value of the determinant of that matrix is zero.

Second Property — If all the rows in a square matrix are interchanged with the corresponding columns, the value of the determinant will not be altered.

Example. Find and compare the determinants of the following square matrices:

$$A = \begin{pmatrix} 3 & 7 & 7 \\ 1 & 2 & 1 \\ 6 & 4 & 5 \end{pmatrix} \qquad B = \begin{pmatrix} 3 & 1 & 6 \\ 7 & 2 & 4 \\ 7 & 1 & 5 \end{pmatrix}$$

NOTE. Matrix B is here said to be *the transpose* of a matrix A, because the columns of matrix A are identical with the rows of matrix B, and *vice versa*.

$$\det A = \begin{pmatrix} 3 & 7 & 7 \\ 1 & 2 & 1 \\ 6 & 4 & 5 \end{pmatrix} \qquad (2\text{-}21)$$

$$(30 + 42 + 28) - (84 + 12 + 35) = (100) - (131) = -31$$

$$\det B = \begin{pmatrix} 3 & 1 & 6 \\ 7 & 2 & 4 \\ 7 & 1 & 5 \end{pmatrix} \qquad (2\text{-}22)$$

$$(30 + 28 + 42) - (84 + 12 + 35) = (100) - (131) = -31$$

Thus $\det A = \det B$.

Third Property — If every element in either a row or a column of a square matrix is multiplied by a constant, the determinant of the new matrix is equal to the product of the constant and the determinant of the original matrix.

Example. Find and compare the determinants of the following square matrices:

$$A = \begin{pmatrix} 3 & 7 & 7 \\ 1 & 2 & 1 \\ 6 & 4 & 5 \end{pmatrix} \qquad B = \begin{pmatrix} 3 & 7 & 7 \\ 2 & 4 & 2 \\ 6 & 4 & 5 \end{pmatrix}$$

NOTE. These two matrices are identical except that the elements in the second row of matrix A have been multiplied by the constant 2 to form the second row of matrix B.

$$\det A = \det \begin{pmatrix} 3 & 7 & 7 \\ 1 & 2 & 1 \\ 6 & 4 & 5 \end{pmatrix} = -31 \text{ (as found for this determinant in the example on Page 51)}$$

$$\det B = \begin{pmatrix} 3 & 7 & 7 \\ 2 & 2 & 4 \\ 6 & 4 & 5 \end{pmatrix} \quad \begin{array}{c} 168 \quad 24 \quad 70 \\ \\ 60 \quad 84 \quad 56 \end{array} \qquad (2\text{-}23)$$

$$(60 + 84 + 56) - (168 + 24 + 70) = (200) - (262) = -62$$

Thus $\det B = 2(\det A)$.

Fourth Property — Any row in a square matrix may be interchanged with any other row without affecting the numerical value of the determinant. However, the sign of the determinant of the newly formed matrix will be the opposite of that of the original determinant. This reversal of sign also occurs when any two columns of a square matrix are interchanged.

Example. Find and compare the determinants of the following square matrices.

$$A = \begin{pmatrix} 3 & 7 & 7 \\ 1 & 2 & 1 \\ 6 & 4 & 5 \end{pmatrix} \qquad B = \begin{pmatrix} 6 & 4 & 5 \\ 1 & 2 & 1 \\ 3 & 7 & 7 \end{pmatrix} \qquad C = \begin{pmatrix} 3 & 7 & 7 \\ 1 & 1 & 2 \\ 6 & 5 & 4 \end{pmatrix}$$

NOTE. Matrix B was formed by interchanging the first and third rows of matrix A. Matrix C was formed by interchanging the second and third columns of matrix A.

As seen in example 1 above, $\det A = -31$.

$$\det B = \begin{pmatrix} 6 & 4 & 5 \\ 1 & 2 & 1 \\ 3 & 7 & 7 \end{pmatrix} \quad \begin{array}{c} 30 \quad 42 \quad 28 \\ \\ 84 \quad 12 \quad 35 \end{array} \qquad (2\text{-}24)$$

$$(84 + 12 + 35) - (30 + 42 + 28) = (131) - (100) = 31$$

Thus det $B = -(\det A)$.

$$\det\ C = \begin{pmatrix} 3 & 7 & 7 & 3 & 7 \\ 1 & 4 & 2 & 1 & 1 \\ 6 & 5 & 4 & 6 & 5 \end{pmatrix} \qquad (2\text{-}25)$$

42 30 28

12 84 35

$(12 + 84 + 35) - (42 + 30 + 28) = (131) - (100) = 31$

Thus det $C = -(\det A)$.

Fifth Property — If any two rows in a square matrix are equal or proportional, the determinant of that matrix is equal to zero. Likewise, if any two columns in a square matrix are equal or proportional, the determinant of that matrix is equal to zero.

Example.   Find the determinants of the following square matrices.

$$A = \begin{pmatrix} 6 & 3 & 5 \\ 6 & 3 & 5 \\ 2 & 1 & 4 \end{pmatrix} \qquad B = \begin{pmatrix} 3 & 1 & 6 \\ 2 & 5 & 4 \\ 4 & 2 & 8 \end{pmatrix}$$

NOTE.   In matrix A rows one and two are equal.   In matrix B columns one and three are in the proportion 1 : 2.

$$\det A = \begin{pmatrix} 6 & 3 & 5 & 6 & 3 \\ 6 & 3 & 5 & 6 & 3 \\ 2 & 1 & 4 & 2 & 1 \end{pmatrix} \qquad (2\text{-}26)$$

30 30 72

72 30 30

$(72 + 30 + 30) - (30 + 30 + 72) = (132) - (132) = 0$

Thus det $A = 0$.

$$\det B = \begin{pmatrix} 3 & 1 & 6 & 3 & 1 \\ 2 & 5 & 4 & 2 & 5 \\ 4 & 2 & 8 & 4 & 2 \end{pmatrix} \qquad (2\text{-}27)$$

120 24 16

120 16 24

$(120 + 16 + 24) - (120 + 24 + 26) = (160) - (160) = 0$

Thus det $B = 0$.

Sixth Property — If in two square matrices having the same dimensions, all rows (or columns) except one are identical, the sum of the two determinants is equal to the value of a determinant having the same elements in the rows (or columns) corresponding to the identical rows (or columns), and elements in the other row (or column) equal to the sums of the elements in the unequal rows (or columns) of the two matrices.

$$= \begin{pmatrix} a & b & c \\ d & e & f \\ g & h & i \end{pmatrix} \qquad B = \begin{pmatrix} a & b & c \\ d' & e' & f' \\ g & h & i \end{pmatrix}$$

then

$$\det \begin{pmatrix} a & b & c \\ d & e & f \\ g & h & i \end{pmatrix} + \det \begin{pmatrix} a & b & c \\ d' & e' & f' \\ g & h & i \end{pmatrix} = \det \begin{pmatrix} a & b & c \\ d+d' & e+e' & f+f' \\ g & h & i \end{pmatrix}$$

(2-28)

This relationship will hold true whenever all but two corresponding rows in a pair of $n \times n$ matrices are equal. The same relationship holds when all but two corresponding columns of a pair of $n \times n$ matrices are equal.

Example. Find the determinants of the following square matrices.

$$A = \begin{pmatrix} 1 & 3 & 4 \\ 4 & 5 & 2 \\ 2 & 1 & 1 \end{pmatrix} \qquad B = \begin{pmatrix} 1 & 3 & 4 \\ 4 & 5 & 2 \\ 3 & 5 & 2 \end{pmatrix} \qquad C = \begin{pmatrix} 1 & 3 & 4 \\ 4 & 5 & 2 \\ 5 & 6 & 3 \end{pmatrix}$$

NOTE. The first and second rows in matrix A are equal to the corresponding rows in matrix B; the third rows differ. The first and second rows in matrix C are the same as the corresponding rows in matrices A and B but the elements in the third row of matrix C are the sums of the corresponding elements in the third rows of matrices A and B.

$$\det A = \begin{pmatrix} 1 & 3 & 4 \\ 4 & 5 & 2 \\ 2 & 1 & 1 \end{pmatrix}$$

(2-29)

$$(5 + 12 + 16) - (40 + 2 + 12) = (33) - (54) = -21$$

$$\det B = \begin{pmatrix} 1 & 3 & 1 & 3 \\ 4 & 5 & 2 & 5 \\ 3 & 6 & 2 & 3 & 5 \end{pmatrix} \quad\quad (2\text{-}30)$$

60 10 24

10 18 80

$$(10 + 18 + 80) - (60 + 10 + 24) = (108) - (94) = 14$$

$$\det C = \begin{pmatrix} 1 & 3 & 4 & 1 & 3 \\ 4 & 5 & 2 & 4 & 5 \\ 5 & 6 & 3 & 5 & 6 \end{pmatrix} \quad\quad (2\text{-}31)$$

100 12 36

15 30 96

$$(15 + 30 + 96) - (100 + 12 + 36) = (141) - (148) = -7$$

Thus:

$$\det A + \det B = \det C$$

Seventh Property — The determinant of a square matrix remains unchanged if to each element of any row is added k times the corresponding element of some other row, where k is any constant. This operation may also be performed on the columns of a matrix without affecting the value of the determinant.

Example. Find and compare the determinants of the following matrices:

$$A = \begin{pmatrix} 1 & 6 & 4 \\ 2 & 1 & 3 \\ 5 & 4 & 2 \end{pmatrix} \quad B = \begin{pmatrix} 1 & 6 & 4 \\ 2 & 1 & 3 \\ 9 & 6 & 8 \end{pmatrix} \quad C = \begin{pmatrix} 1 & 7 & 4 \\ 2 & 3 & 3 \\ 5 & 9 & 2 \end{pmatrix}$$

NOTE. The first two rows of matrices A and B are identical but the last row of B has been found by adding to each element in the third row of A twice the value of each element in the second row of A. Thus $5 + 2(2) = 9$, etc. Likewise the first and the third columns in matrices $A + C$ are identical but the second column in matrix C has been found by adding the elements of the first and second columns of matrix A. Thus $1 + 6 = 7$, etc.

$$\det A = \begin{pmatrix} 1 & 6 & 4 \\ 2 & 3 & 1 \\ 5 & 2 & 4 \end{pmatrix} \qquad (2\text{-}32)$$

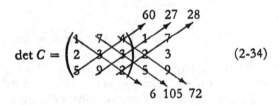

$$(2 + 90 + 32) - (20 + 12 + 24) = (124) - (56) = 68$$

$$\det B = \begin{pmatrix} 1 & 6 & 4 \\ 2 & 3 & 1 \\ 9 & 6 & 6 \end{pmatrix} \qquad (2\text{-}33)$$

$$(8 + 162 + 48) - (36 + 18 + 96) = (218) - (160) = 68$$

$$\det C = \begin{pmatrix} 1 & 7 & 4 \\ 2 & 3 & 3 \\ 5 & 9 & 2 \end{pmatrix} \qquad (2\text{-}34)$$

$$(6 + 105 + 72) - (60 + 27 + 28) = (183) - (115) = 68$$

Thus $\det A = \det B = \det C$.

32. Minors and the Evaluation of 4th Order Determinants. We have already shown that the expanded value of a third order determinant is the algebraic sum of the six signed products obtained (Page 49).

$$\det A = \det \begin{pmatrix} a_1 & a_2 & a_3 \\ b_1 & b_2 & b_3 \\ c_1 & c_2 & c_3 \end{pmatrix}$$

$$= a_1 b_2 c_3 + a_2 b_3 c_1 + a_3 b_1 c_2 - a_1 b_3 c_2 - a_2 b_1 c_3 - a_3 b_2 c_1$$

A third order determinant can also be expressed as the sum of three second order determinants when each of these has been multiplied by an element of a row or column in the original determinant. If the above determinant were expressed by this method, it would appear as:

$$\det A = \det \begin{pmatrix} a_1 & a_2 & a_3 \\ b_1 & b_2 & b_3 \\ c_1 & c_2 & c_3 \end{pmatrix}$$

$$= (a_1) \det \begin{pmatrix} b_2 & b_3 \\ c_2 & c_3 \end{pmatrix} - (a_2) \det \begin{pmatrix} b_1 & b_3 \\ c_1 & c_3 \end{pmatrix} + (a_3) \det \begin{pmatrix} b_1 & b_2 \\ c_1 & c_2 \end{pmatrix}$$

$$(2\text{-}35)$$

This process and its resulting expression are described as the expansion of $\det A$ by minors according to the elements in its first row. The *minor* of the element in the pth row and qth column of a determinant of any order n is a determinant of the order $n - 1$ produced by discarding all the elements in the pth row and qth column in the original matrix. Thus the minor (A_1) of the element a_1 in matrix A above was obtained as shown below:

$$A_1 = \begin{pmatrix} a_1 & a_2 & a_3 \\ b_1 & b_2 & b_3 \\ c_1 & c_2 & c_3 \end{pmatrix} = b_2 c_3 - b_3 c_2$$

Likewise the minors $(A_2 \text{ and } A_3)$ of the elements a_2 and a_3 in matrix A were obtained by extending this operation.

$$A_2 = \begin{pmatrix} a_1 & a_2 & a_3 \\ b_1 & b_2 & b_3 \\ c_1 & c_2 & c_3 \end{pmatrix} = b_1 c_3 - b_3 c_1$$

$$A_3 = \begin{pmatrix} a_1 & a_2 & a_3 \\ b_1 & b_2 & b_3 \\ c_1 & c_2 & c_3 \end{pmatrix} = b_1 c_2 - b_2 c_1$$

Notice above that when these minors are multiplied by their respective elements and added, the second term in the sum is preceded by a minus sign. A minor that is prefixed by a definite sign is called a cofactor. The sign of the cofactor of an element in the pth row and qth column of a matrix is found by applying the following equation:

$$\text{cofactor} = (-1)^{p+q} \text{(minor)} \qquad (2\text{-}36)$$

Thus in general we may say that the determinant of a matrix can be expressed as the sum of the products of the elements of any row or

column by their corresponding cofactors. We now use this technique
to find the determinants of square matrices of the fourth order.

Example. Find the determinant of matrix A below.

$$\det A = \det \begin{pmatrix} 4 & 1 & 2 & 1 \\ 2 & 3 & 1 & 7 \\ 5 & 4 & 10 & 2 \\ 3 & 4 & 5 & 1 \end{pmatrix}$$

First, we know that we must expect the desired sum to have $n! = 1 \cdot 2 \cdot 3 \cdot 4 = 24$ terms and that each term must be the product of four elements from the given matrix. We also know that no two elements in any one term may come from the same row or column of the given matrix. It can easily be seen that the method of repeating columns to the right of the given matrix and then taking products along diagonals, as we did in the solution of 3×3 matrices, is not applicable to the solution of 4×4 matrices. So let us now use cofactors to obtain the determinant of this matrix.

Thus:

$$\det A = + (4) \det \begin{pmatrix} 3 & 1 & 7 \\ 4 & 10 & 2 \\ 4 & 5 & 1 \end{pmatrix} - (1) \det \begin{pmatrix} 2 & 1 & 7 \\ 5 & 10 & 2 \\ 3 & 5 & 1 \end{pmatrix}$$

$$+ (2) \det \begin{pmatrix} 2 & 3 & 7 \\ 5 & 4 & 2 \\ 3 & 4 & 1 \end{pmatrix} - (1) \det \begin{pmatrix} 2 & 3 & 1 \\ 5 & 4 & 10 \\ 3 & 4 & 5 \end{pmatrix}$$

Here we see that a fourth order determinant may be expressed by the sum of the products of the elements in the first row of the given matrix multiplied by their respective cofactors. The determinants of these cofactors can be found easily by applying the methods used in obtaining the determinant of a 3×3 matrix.

$$\begin{aligned}
\det A = &+ (4)(30 + 140 + 8 - 280 - 30 - 4) \\
&- (1)(20 + 175 + 6 - 210 - 20 - 5) \\
&+ (2)(8 + 140 + 18 - 84 - 16 - 15) \\
&- (1)(40 + 20 + 90 - 12 - 80 - 75)
\end{aligned}$$

$$\det A = (4)(-136) - (1)(-34) + (2)(51) - (1)(-17)$$

$$\det A = -544 + 34 + 102 + 17 = -391$$

This operation for finding the determinant of a 4×4 matrix may be represented symbolically as follows:

$$\det A = \det \begin{pmatrix} a_1 & a_2 & a_3 & a_4 \\ b_1 & b_2 & b_3 & b_4 \\ c_1 & c_2 & c_3 & c_4 \\ d_1 & d_2 & d_3 & d_4 \end{pmatrix} = (a_1) \det \begin{pmatrix} b_2 & b_3 & b_4 \\ c_2 & c_3 & c_4 \\ d_2 & d_3 & d_4 \end{pmatrix}$$

$$- (a_2) \det \begin{pmatrix} b_1 & b_3 & b_4 \\ c_1 & c_3 & c_4 \\ d_1 & d_3 & d_4 \end{pmatrix} + (a_3) \det \begin{pmatrix} b_1 & b_2 & b_4 \\ c_1 & c_2 & c_4 \\ d_1 & d_2 & d_4 \end{pmatrix}$$

$$- (a_4) \det \begin{pmatrix} b_1 & b_2 & b_3 \\ c_1 & c_2 & c_3 \\ d_1 & d_2 & d_3 \end{pmatrix} \qquad (2\text{-}37)$$

Or $\det A = a_1 \det A_1 - a_2 \det A_2 + a_3 \det A_3 - a_4 \det A_4$ where $a_1 \cdots$ represent elements in the first row of the given matrix and $A_1 \cdots$ represent their corresponding minors.

Or $\det A = a_1 \det A'_1 + a_2 \det A'_2 + a_3 \det A'_3 + a_4 \det A'_4$ where $A'_1 \cdots$ here represent the cofactors corresponding to the given elements.

33. The Solution of Simultaneous Linear Equations by Determinants. One of the most practical applications of determinants arises in the solution of simultaneous linear equations. In example 1 below we recall the algebraic method of multiplication and subtraction for solving a pair of simultaneous linear equations in two unknowns.

Example 1.   Solve the following linear equations:

$$4x - 3y = 6$$

$$3x - 2y = 5$$

By multiplying the first equation through by $(+2)$ and the second equation through by $(+3)$, we obtain:

$$8x - 6y = 12$$

$$9x - 6y = 15$$

Then, if we subtract the first equation from the second we obtain

$$x = 3$$

In order to find y, we return to our original equations and multiply the first through by $(+3)$ and the second through by $(+4)$. Then we have

$$12x - 9y = 18$$

$$12x - 8y = 20$$

When we subtract the first equation from the second we find that

$$y = 2$$

Let us carry out this solution again by using literal algebraic numbers for the coefficients of x and y, and for the constant terms.

$$a_1x + b_1y = k_1 \qquad (2\text{-}38)$$

$$a_2x + b_2y = k_2 \qquad (2\text{-}39)$$

a_1 and a_2 represent the coefficients of x; b_1 and b_2, those of y; and k_1 and k_2, the constant terms. In order to find the value of x our first step is to multiply both sides of equation (2-38) by b_2 and both sides of equation (2-39) by b_1. This gives us:

$$b_2a_1x + b_2b_1y = b_2k_1 \qquad (2\text{-}40)$$

$$b_1a_2x + b_1b_2y = b_1k_2 \qquad (2\text{-}41)$$

If we now subtract equation (2-41) from equation (2-40), we find that the y term disappears and that we are left with

$$b_2a_1x - b_1a_2x = b_2k_1 - b_1k_2$$

which becomes, by factoring the left-hand member of the equation.

$$(b_2a_1 - b_1a_2)\, x = b_2k_1 - b_1k_2$$

Then by transposition we obtain

$$x = \frac{b_2k_1 - b_1k_2}{b_2a_1 - b_1a_2} \quad \text{or} \quad \frac{b_2k_1 - b_1k_2}{a_1b_2 - a_2b_1} \qquad (2\text{-}42)$$

If the terms are now eliminated by multiplying both sides of equation (2-38) by a_2 and both sides of equation (2-39) by a_1, and subtracting, we obtain

$$v = \frac{a_1k_2 - a_2k_1}{a_1b_2 - a_2b_1} \qquad (2\text{-}43)$$

If we now compare equation (2-42) with equation (2-43), we notice that their denominators are identical.

Furthermore, this common denominator is seen to be the cross-product of the coefficients on the left-hand side of the equations. As such it may be represented by the following determinant*

$$\begin{vmatrix} a_1 & b_1 \\ a_2 & b_2 \end{vmatrix} = (a_1 b_2 - a_2 b_1) \tag{2-44}$$

Likewise the numerator of equation (2-42) may be represented as

$$\begin{vmatrix} k_1 & b_1 \\ k_2 & b_2 \end{vmatrix} = (b_2 k_1 - b_1 k_2) \tag{2-45}$$

and that of equation (2-43) as

$$\begin{vmatrix} a_1 & k_1 \\ a_2 & k_2 \end{vmatrix} = (a_1 k_2 - a_2 k_1) \tag{2-46}$$

Thus, one way of solving such a pair of simultaneous equations is to represent x and y as the quotients of two determinants.

$$x = \frac{\begin{vmatrix} k_1 & b_1 \\ k_2 & b_2 \end{vmatrix}}{\begin{vmatrix} a_1 & b_1 \\ a_2 & b_2 \end{vmatrix}} \tag{2-47} \qquad\qquad y = \frac{\begin{vmatrix} a_1 & k_1 \\ a_2 & k_2 \end{vmatrix}}{\begin{vmatrix} a_1 & b_1 \\ a_2 & b_2 \end{vmatrix}} \tag{2-48}$$

Example. Solve the following linear equations with the aid of determinants

$$4x - 3y = 6$$
$$3x - 2y = 5$$

$$x = \frac{\begin{vmatrix} 6 & -3 \\ 5 & -2 \end{vmatrix}}{\begin{vmatrix} 4 & -3 \\ 3 & -2 \end{vmatrix}} = \frac{(6)(-2) - (-3)(5)}{(4)(-2) - (-3)(3)} = \frac{-12 + 15}{-8 + 9} = 3 \tag{2-49}$$

$$y = \frac{\begin{vmatrix} 4 & 6 \\ 3 & 5 \end{vmatrix}}{\begin{vmatrix} 4 & -3 \\ 3 & -2 \end{vmatrix}} = \frac{(4)(5) - (6)(3)}{(4)(-2) - (-3)(3)} = \frac{20 - 18}{-8 + 9} = 2 \tag{2-50}$$

*In the earlier part of this chapter, determinants were denoted by placing the letters "det" before their matrices, which were expressed by the use of parentheses. When determinants appear without reference to their matrices, their elements are enclosed in vertical lines, as stated at the beginning of this chapter.

These values for x and y are seen to be identical with those found in the example on Page 59 for the same equations.

The advantages of using determinants for the solution of linear equations are more apparent in the following example:

Example 2. Solve the following pair of linear equations by making use of determinants.

$$2x + 1.5y = 18.45$$

$$5x - 4y = 12.8$$

By equation (2-47) we have

$$x = \frac{\begin{vmatrix} 18.45 & 1.5 \\ 12.8 & -4 \end{vmatrix}}{\begin{vmatrix} 2 & 1.5 \\ 5 & -4 \end{vmatrix}} = \frac{-73.8 - 19.2}{-8 - 7.5} = \frac{-93}{-15.5}$$

Thus $x = 6$

While by equation (2-48) we have

$$y = \frac{\begin{vmatrix} 2 & 18.45 \\ 5 & 12.8 \end{vmatrix}}{\begin{vmatrix} 2 & 1.5 \\ 5 & -4 \end{vmatrix}} = \frac{25.6 - 92.25}{-15.5*} = \frac{-66.65}{-15.5}$$

Thus $y = 4.3$

The method of this operation can easily be expanded to facilitate the solution of three simultaneous linear equations in three unknowns. Because of the similarities between the derivation of this method and that just explained for equations in two unknowns, only the results are given below.

Given the following linear equations in three variables.

$$a_1x + b_1y + c_1z = k_1$$

$$a_2x + b_2y + c_2z = k_2$$

$$a_3x + b_3y + c_3z = k_3$$

The solution by determinants yields

$$x = \frac{\begin{vmatrix} k_1 & b_1 & c_1 \\ k_2 & b_2 & c_2 \\ k_3 & b_3 & c_3 \end{vmatrix}}{\begin{vmatrix} a_1 & b_1 & c_1 \\ a_2 & b_2 & c_2 \\ a_3 & b_3 & c_3 \end{vmatrix}}; \quad y = \frac{\begin{vmatrix} a_1 & k_1 & c_1 \\ a_2 & k_2 & c_2 \\ a_3 & k_3 & c_3 \end{vmatrix}}{\begin{vmatrix} a_1 & b_1 & c_1 \\ a_2 & b_2 & c_2 \\ a_3 & b_3 & c_3 \end{vmatrix}}; \quad z = \frac{\begin{vmatrix} a_1 & b_1 & k_1 \\ a_2 & b_2 & k_2 \\ a_3 & b_3 & k_3 \end{vmatrix}}{\begin{vmatrix} a_1 & b_1 & c_1 \\ a_2 & b_2 & c_2 \\ a_3 & b_3 & c_3 \end{vmatrix}} \qquad (2\text{-}51)$$

*As stated above, the values of the two denominators are identical.

Example 3. Solve the following linear equations by the use of determinants

$$2x - y + z = 10$$

$$x + 3y - 2z = -4$$

$$5x + 2y - 3z = 2$$

Then

$$x = \frac{\begin{vmatrix} 10 & -1 & 1 \\ -4 & 3 & -2 \\ 2 & 2 & -3 \end{vmatrix}}{\begin{vmatrix} 2 & -1 & 1 \\ 1 & 3 & -2 \\ 5 & 2 & -3 \end{vmatrix}} \tag{2-52}$$

$$= \frac{-90 + 4 - 8 + 40 + 12 - 6}{-18 + 10 + 2 + 8 - 3 - 15} = \frac{-48}{-16} = 3$$

$$y = \frac{\begin{vmatrix} 2 & 10 & 1 \\ 1 & -4 & -2 \\ 5 & 2 & -3 \end{vmatrix}}{-16} \tag{2-53}$$

$$= \frac{24 - 100 + 2 + 8 + 30 + 20}{-16} = \frac{-16}{-16} = 1$$

$$z = \frac{\begin{vmatrix} 2 & -1 & 10 \\ 1 & 3 & -4 \\ 5 & 2 & 2 \end{vmatrix}}{-16} \tag{2-54}$$

$$= \frac{12 + 20 + 20 + 16 + 2 - 150}{-16} = \frac{-80}{-16} = 5$$

Thus $x = 3$; $y = 1$; $z = 5$

There is one case for which determinants will not be able to provide unique solutions for a similar set of linear equations. This occurs when the determinant of the matrix serving as the common denominator is equal to zero.

Example 4. Solve the following set of linear equations.

$$3x + 2y + 5z = 14 \tag{2-55}$$

$$x + 3y + 4z = 14 \tag{2-56}$$

$$2x + 5y + 7z = 24 \tag{2-57}$$

$$\text{Then } x = \frac{\begin{vmatrix} 14 & 2 & 5 \\ 14 & 3 & 4 \\ 24 & 5 & 7 \end{vmatrix}}{\begin{vmatrix} 3 & 2 & 5 \\ 1 & 3 & 4 \\ 2 & 5 & 7 \end{vmatrix}} = \frac{294 + 192 + 350 - 280 - 196 - 360}{63 + 16 + 25 - 60 - 14 - 30} = \frac{0}{0} \tag{2-58}$$

This case corresponds to the result found by the algebraic method of solution when elimination of a variable from two pairs of equations gave the same equations in two unknowns. To show this, let us eliminate the term x from equations (2-55) and (2-56), and (2-56) and (2-57) above: thus:

$$3x + 9y + 12z = 42$$
$$\underline{-(3x + 2y + 5z = 14)}$$
$$7y + 7z = 28$$

dividing through by 7, we have

$$y + z = 4 \tag{2-59}$$

Similarly

$$2x + 6y + 8z = 28$$
$$\underline{-(2x + 5y + 7z = 24)}$$
$$y + z = 4$$

Since the results of the elimination of x from equations (2-55) and (2-56), and from equations (2-56) and (2-57) are the same equation in y and z, we do not have the necessary two independent equations to solve for those two unknowns. Moreover, it follows that elimination of y or z instead of x from any two pairs of the three equations would have the same indeterminate result. Therefore, the closest approach we can make to a solution is to express two of the variables in terms of the third, as from the equation found above, we have $z = 4 - y$, and by eliminating z from the original equations we would also have $x = y - 2$. This situation, and the use of determinants (and matrices) in investigating it, is discussed in the next section.

34. Rank of a Matrix and Consistency of Equations. An important use of matrices and determinants is in investigating the consistency of two or more simultaneous algebraic equations. Consistency is the property possessed by equations when they are all satisfied by at least one set of values of the variables, i.e., their loci all have one or more common points.   If they are not satisfied by any one set of values of the variables, they are said to be inconsistent.   E.g., the equations $x + y = 4$ and $x + y = 5$ are inconsistent.   The equations $x + y - 4$ and $2x + 2y = 8$ are consistent, but are not independent: and the equations $x + y = 4$ and $x - y = 2$ are consistent and independent.   The first pair of equations represents two parallel lines, the second represents two coincident lines, and the third represents two distinct lines intersecting in a point, the point whose coordinates are (3, 1).

A linear equation in two variables is the equation of a line in the plane.   Therefore a single equation in two variables has an unlimited number of solutions.   Two equations have a unique simultaneous solution if the lines they represent intersect and are not coincident; there is no solution if the lines are parallel and not coincident; there is an unlimited number of solutions if the lines are coincident.   These correspond to the three cases of the following discussion.   Consider the equations: $a_1x + b_1y = k_1$, $a_2x + b_2y = k_2$, where at least one of a_1, b_1 and at least one of a_2, b_2 is not zero.   As explained earlier in this chapter multiply the first equation by b_2 and the second by b_1, then subtract. This gives $(a_1b_2 - a_2b_1)x = b_2k_1 - b_1k_2$. Similarly, $(a_1b_2 - a_2b_1)y = a_1k_2 - a_2k_1$,

or

$$x \begin{vmatrix} a_1 & b_1 \\ a_2 & b_2 \end{vmatrix} = \begin{vmatrix} k_1 & b_1 \\ k_2 & b_2 \end{vmatrix}, \tag{2-61}$$

and

$$y \begin{vmatrix} a_1 & b_1 \\ a_2 & b_2 \end{vmatrix} = \begin{vmatrix} a_1 & k_1 \\ a_2 & k_2 \end{vmatrix}.$$

Three cases follow: I.   If the determinant of the coefficients

$$\begin{vmatrix} a_1 & b_1 \\ a_2 & b_2 \end{vmatrix}$$

is not zero, one can divide by it and secure unique values for x and y. The equations are then consistent and independent.   The equations $2x - y = 1$ and $x + y = 3$ reduce in the above way to

$$3x = 4 \quad \text{and} \quad 3y = 5$$

and have the unique simultaneous solution $x = \frac{4}{3}, y = \frac{5}{3}$. II. If the determinant of the coefficients is zero and one of the determinants formed by replacing the coefficients of x (or of y) by the constant terms is not zero, there is no solution; i.e., the equations are inconsistent. The equations $2x - y = 1$ and $4x - 2y = 3$ reduce to

$$0 \cdot x = 1 \quad \text{and} \quad 0 \cdot y = 2,$$

which have no solution. III. If all three determinants are zero, there results $0 \cdot x = 0$ and $0 \cdot y = 0$. The equations are then consistent but not independent. This is the situation for the equations $x - y = 1$ and $2x - 2y = 2$. An infinite number of pairs of values of x and y can be found that satisfy both of these equations.

A linear equation in three variables is the equation of a plane in space. Therefore a single equation has an unlimited number of solutions. Two equations either represent parallel planes and have no common solution or else represent planes which intersect in a line, or coincide so the equations have an unlimited number of solutions. As was shown in this chapter, eliminating the variables, two at a time, from the equations

$$a_1 x + b_1 y + c_1 z = k_1,$$

$$a_2 x + b_2 y + c_2 z = k_2, \qquad (2\text{-}62)$$

$$a_3 x + b_3 y + c_3 z = k_3,$$

gives $Dx = K_1$, $Dy = K_2$, and $Dz = K_3$, where K_1, K_2, and K_3 are the determinants resulting from substituting the k's in the determinant of the coefficients, D, in place of the a's, b's, and c's, respectively. Three cases arise: I. If $D \neq 0$, it can be divided out and a unique set of values for x, y, and z obtained; i.e., the three planes representing the three equations then intersect in a point and the equations are consistent (and also independent). II. If $D = 0$ and at least one of K_1, K_2, and K_3 is not zero, there is no solution; the three planes do not have any point in common and the three equations are inconsistent. III. If $D = 0$ and $K_1 = K_2 = K_3 = 0$, three cases arise: a). Some second-order determinant in D is not zero, in which case the equations have infinitely many points in common; the planes (the loci of the equations) intersect in a line and the equations are consistent. b). Every second-order minor in D is zero and a second-order minor in K_1,

K_2, or K_3 is not zero.   The planes are then parallel, but at least one pair do not coincide; the equations are inconsistent.   c).   All second-order minors in D, K_1, K_2 and K_3 are zero.   The three planes then coincide and the equations are consistent (but not independent).

The general situation of m linear equations in n unknowns is best handled by consideration of matrix rank.   The rank of a matrix is the order of the non-zero determinant of greatest order than can be selected from the matrix by taking out rows and columns.   The concept of rank thus facilitates the statement of the condition for consistency of simultaneous linear equations: m linear equations in n unknowns are consistent when, and only when, the rank of the matrix of the coefficients is equal to the rank of the augmented matrix.   In the system of linear equations

$$\left. \begin{array}{l} x + y + z + 3 = 0 \\ 2x + y + z + 4 = 0, \end{array} \right\} \tag{2-63}$$

the matrix of the coefficients is

$$\begin{pmatrix} 1 & 1 & 1 \\ 2 & 1 & 1 \end{pmatrix}$$

and the augmented matrix is

$$\begin{pmatrix} 1 & 1 & 1 & 3 \\ 2 & 1 & 1 & 4 \end{pmatrix}.$$

The rank of both is two, because the determinant

$$\begin{vmatrix} 1 & 1 \\ 2 & 1 \end{vmatrix}$$

is not zero.   Hence these equations are satisfied by some set of values of x and y and z.

If the constant terms in a system of linear equations are all zero (the equations are homogeneous), then the equations have a trivial solution (each unknown equal to zero).   For n homogeneous linear equations in m unknowns: (1) If $n < m$, the equations have a nontrivial solution (not all unknowns zero).   (2) If $n = m$, the equations have a nontrivial solution if, and only if, the determinent of the coefficients is equal to zero.   (3) If $n < m$, the equations have a nontrivial solution if, and only if, the rank of the matrix of the coefficients is less than m.

These are simply the special case of the results for n linear equations in m unknowns when the constant terms are all zero.

35. Characteristic Equation of a Matrix. Let I be the unit matrix of the same order as the square matrix A. Then compute the product matrix of I by x, and subtract the matrix A from the matrix xI, obtaining a difference matrix $xI - A$. (The unit matrix has 1 for all elements along the main diagonal, all other elements being zero).* The characteristic equation of the matrix A is then the determinant of the matrix $xI - A$, equated to 0.

For example, find the characteristic equation of the matrix

$$A = \begin{pmatrix} 1 & 3 \\ 2 & 4 \end{pmatrix} \tag{2-64}$$

Now the 2 × 2 unit matrix I is

$$I = \begin{pmatrix} 1 & 0 \\ 0 & 1 \end{pmatrix} \tag{2-65}$$

as defined above. Then the matrix xI is

$$xI = x\begin{pmatrix} 1 & 0 \\ 0 & 1 \end{pmatrix} = \begin{pmatrix} x & 0 \\ 0 & x \end{pmatrix} \tag{2-66}$$

Now the matrix $-A$ is

$$-A = \begin{pmatrix} -1 & -3 \\ -2 & -4 \end{pmatrix} \tag{2-67}$$

Therefore $xI - A$ is

$$\begin{pmatrix} x & 0 \\ 0 & x \end{pmatrix} + \begin{pmatrix} -1 & -3 \\ -2 & -4 \end{pmatrix} = \begin{pmatrix} x-1 & -3 \\ -2 & x-4 \end{pmatrix} \tag{2-68}$$

and the determinant of this matrix is

$$\det \begin{pmatrix} x-1 & -3 \\ -2 & x-4 \end{pmatrix} = \begin{vmatrix} x-1 & -3 \\ -2 & x-4 \end{vmatrix} = 0 \tag{2-69}$$

$$(x-1)(x-4) - ((-3)(-2)) = 0$$

$$x^2 - 5x + 4 - 6 = x^2 - 5x - 2 = 0.$$

Therefore the characteristic equation of matrix A is $x^2 - 5x - 2 = 0$.

$$\tag{2-70}$$

*See Section 38, p. 75.

An important property of all square matrices is that they satisfy their characteristic equations. Thus, if one were to perform the operations of multiplication and addition on matrix A corresponding to the above characteristic equation, that is,

$$(A \cdot A) - (5 \cdot A) - (2 \cdot I)$$

he would find that the result was zero.

The root of the characteristic equation of a square matrix is called the *characteristic root of the matrix*. This root is also called a *proper value*, an *eigenvalue*, a *latent root* or a *characteristic number* of its square matrix.

As a physical example, consider a suspension bridge which is vibrating with n degrees of freedom about a position of stable equilibrium. Its potential energy at any time will be a function, call it $A(q_1, q_2, \cdots, q_n)$, of its n position coordinates $q_1, q_2, \cdots, q_n$, which is approximated by the quadratic form $\sum a_{ik} q_i q_k$, the constants a_{ik} being given by $a_{ik} = \frac{1}{2}(\partial^2 A / \partial q_i \partial q_k)$ evaluated at the position of equilibrium. Then, the characteristic roots of the matrix (a_{ik}) are the squares of the critical frequencies of the bridge, namely the frequencies at which an impressed force will cause dangerous resonance.

36. Vectors. While a vector is defined as a quantity having magnitude and direction, it has equivalent mathematical definitions, one of which is simply a linear array of numbers. Therefore a matrix that has only one row is often called a *row vector*, as is any one of the rows in a matrix that has more than one row. Similarly a matrix that has only one column is called a *column vector*, as is any column of a matrix that has more than one column. A *proper vector*, or *eigenvector*, is a non-zero vector **x** such that its product by a matrix A, which is $A\mathbf{x}$, is equal to its product by the characteristic root of the matrix, λ, which is $\lambda\mathbf{x}$.

37. The Application of Matrices in Describing Geometrical Distortions and Transformations. The application of determinants in solving simultaneous linear equations provides one example of the usefulness of matrix notation. Another important application of this notation occurs in the study of geometrical distortions. The first distortion of this type that we will consider is called a *stretch*. In Figure 2-1 the line OP_o is stretched to form the line OP_1. In other words, P_o travels to the point P_1 on the line OP_o and the distance OP_1 is now described as being $(k)OP_o$ where k is a constant. This operation then causes a diagram to change its scale by a factor k. In the following figures capital letters followed by subscripts will be used to denote points and small letters followed by subscripts will be used to denote the coordinates of a point. Small letters without subscripts will indicate distances or constants.

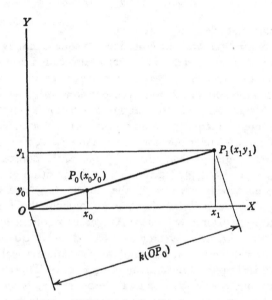

Fig. 2-1. The stretch of a line segment from
OP_0 to OP_1.

According to Figure 2-1, the distance OP_1 is seen to be equal to $k(\overline{OP_0})$. This stretch was effected by multiplying each of the coordinates of P_0 by the constant k. Thus we may represent this type of distortion by the following set of equations.

$$\left. \begin{array}{l} x_1 = kx_0 \\ y_1 = ky_0 \end{array} \right\} \tag{2-71}$$

It is a simple operation to convert these expressions to matrix form. We then have

$$\begin{pmatrix} x_1 \\ y_1 \end{pmatrix} = \begin{pmatrix} k & 0 \\ 0 & k \end{pmatrix} \begin{pmatrix} x_0 \\ y_0 \end{pmatrix} \tag{2-72}$$

When the right-hand side of this equation undergoes the operation of row by column multiplication we find that

$$\begin{pmatrix} x_1 \\ y_1 \end{pmatrix} = \begin{pmatrix} k & 0 \\ 0 & k \end{pmatrix} \begin{pmatrix} x_o \\ y_o \end{pmatrix} = \begin{pmatrix} kx_o \\ ky_o \end{pmatrix} \qquad (2\text{-}73)$$

The matrix $\begin{pmatrix} k & 0 \\ 0 & k \end{pmatrix}$ is called an *operator*, for it is used to transform the line OP_o into OP_1. It is also interesting to note that the area of triangle OP_ox_o is related to the area of triangle OP_1x_1 by multiplication by a factor equal to the determinant of the operator:

$$\det \begin{pmatrix} k & 0 \\ 0 & k \end{pmatrix} = k^2 \qquad \text{Area } \Delta OP_1x_1 = (k^2)\text{ area } \Delta OP_ox_o \qquad (2\text{-}74)$$

Example. Using Figure 2-1, if the coordinates of P_o are (4,3) and those of P_1 are (24, 18), what is the matrix equation that describes this stretch and by how much do the areas of OP_1x_1 and OP_ox_0 differ?

By simple calculation and substitution we obtain

$$\begin{pmatrix} 24 \\ 18 \end{pmatrix} = \begin{pmatrix} 6 & 0 \\ 0 & 6 \end{pmatrix} \begin{pmatrix} 4 \\ 3 \end{pmatrix}$$

The determinant of the operator is 36 so we immediately know that the area of ΔOP_1x_1 is equal to (36)·area ΔOP_ox_o.

$$\text{Area } \Delta OP_ox_0 = \frac{4 \times 3}{2} = 6$$

Therefore, $\text{Area } \Delta OP_1x_1 = 36 \times 6 = 216 \qquad (2\text{-}75)$

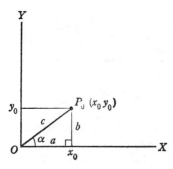

FIG. 2-2a.   Triangle Ox_0P_0 in its original position.

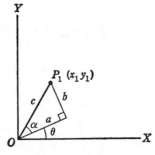

FIG. 2-2b.   The same triangle after being rotated through an angle θ.

Another process of geometrical distortion that can be expressed by matrices is called a *rotation*. Consider the triangle Ox_0P_0 in Figure 2-2a. If we wish to express the rotation of this triangle through an angle θ to the position shown in Figure 2-2b, we must first find an expression of the relationship between the coordinates of P_0 and P_1.

By applying simple trigonometry we find that

$$\left.\begin{array}{l} x_1 = x_0 \cos\theta - y_0 \sin\theta \\[4pt] y_1 = x_0 \sin\theta + y_0 \cos\theta \end{array}\right\} \tag{2-76}$$

which can be expressed in matrix notation as

$$\begin{pmatrix} x_1 \\ y_1 \end{pmatrix} = \begin{pmatrix} \cos\theta & -\sin\theta \\ \sin\theta & +\cos\theta \end{pmatrix}\begin{pmatrix} x_0 \\ y_0 \end{pmatrix} \tag{2-77}$$

The square matrix here is again referred to as an operator. Its determinant is seen to be $\cos^2\theta + \sin^2\theta$ which is equal to 1 for any value of θ. Since the product of the determinant of the operator and the area of the first triangle is equal to the area of the transformed triangle, this unitary value of the operator expresses the fact that the area is unchanged by rotation.

In still another transformation process, the triangle OP_0x_0 may be reflected as illustrated in Figure 2-3.

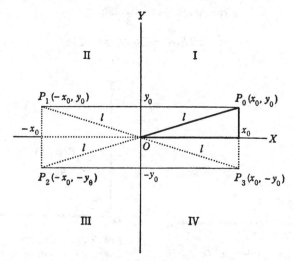

Fig. 2-3.　Triangle OP_0x_0 and its three reflections.

Here we see that the point P_0, lying in the first quadrant (denoted by the roman numeral I) has a reflection in each of the other quadrants of the XY-plane. The coordinates of the reflected points are described by the following set of matrix equations

$$P_1 = \begin{pmatrix} -1 & 0 \\ 0 & 1 \end{pmatrix} \begin{pmatrix} x_0 \\ y_0 \end{pmatrix} = \begin{pmatrix} -x_0 \\ y_0 \end{pmatrix}$$

$$P_2 = \begin{pmatrix} -1 & 0 \\ 0 & -1 \end{pmatrix} \begin{pmatrix} x_0 \\ y_0 \end{pmatrix} = \begin{pmatrix} -x_0 \\ -y_0 \end{pmatrix} \qquad (2\text{-}78)$$

$$P_3 = \begin{pmatrix} 1 & 0 \\ 0 & -1 \end{pmatrix} \begin{pmatrix} x_0 \\ y_0 \end{pmatrix} = \begin{pmatrix} x_0 \\ -y_0 \end{pmatrix}$$

Thus we see that the operators

$$\begin{pmatrix} -1 & 0 \\ 0 & 1 \end{pmatrix}, \qquad \begin{pmatrix} -1 & 0 \\ 0 & -1 \end{pmatrix} \quad \text{and} \quad \begin{pmatrix} 1 & 0 \\ 0 & -1 \end{pmatrix}$$

can be used to describe the reflections in the second, third and fourth quadrants of a line or a triangle in the first quadrant.

The determinants of these three matrices are -1, $+1$, and -1 respectively. The absolute value of each of these determinants is 1 and this fact indicates that there is no difference in the size of the four triangles or that

$$\triangle x_0 O P_0 = \triangle -x_0 O P_1 = \triangle -x_0 O P_2 = \triangle x_0 O P_3, \text{ in area.} \qquad (2\text{-}79)$$

But the signs of these determinants do have an important significance in describing the position of the reflected triangle. We can see that $\triangle x_0 O P_0$ when rotated by 180°, is congruent with $\triangle -x_0 O P_2$. This is consistent with the fact that the determinant of the operator used for this reflection is $+1$. However, we can not make $\triangle x_0 O P_0$ coincide with either $\triangle -x_0 O P_1$ or $\triangle x_0 O P_3$ by rotation alone. In order to make the original triangle coincide with these two reflected triangles, we must flip the original triangle over. When this has been done, the original triangle can be made to coincide with its reflections in the second and fourth quadrants. Thus the minus signs prefixing the determinants of these operators indicate that the original triangle must be inverted before it can be made to coincide with its reflections in the second and fourth quadrants.

A matrix may also be used to transform a given rectangle into a parallelogram. Consider the rectangle in Figure 2-4 defined by the points $P_0 = (0, 0)$, $P_1 = (5, 0)$, $P_2 = (5, 3)$, $P_3 = (0, 3)$.

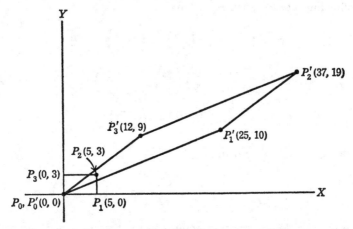

Fig. 2-4.   The transformation of a rectangle into a parallelogram.

Let us see what occurs when we operate on this figure with the matrix

$$\begin{pmatrix} 5 & 4 \\ 2 & 3 \end{pmatrix}$$

Then we have

$$P'_0 = \begin{pmatrix} 5 & 4 \\ 2 & 3 \end{pmatrix}\begin{pmatrix} 0 \\ 0 \end{pmatrix} = \begin{pmatrix} 0 \\ 0 \end{pmatrix}$$

$$P'_1 = \begin{pmatrix} 5 & 4 \\ 2 & 3 \end{pmatrix}\begin{pmatrix} 5 \\ 0 \end{pmatrix} = \begin{pmatrix} 25 \\ 10 \end{pmatrix}$$

$$P'_2 = \begin{pmatrix} 5 & 4 \\ 2 & 3 \end{pmatrix}\begin{pmatrix} 5 \\ 3 \end{pmatrix} = \begin{pmatrix} 37 \\ 19 \end{pmatrix}$$

$$P'_3 = \begin{pmatrix} 5 & 4 \\ 2 & 3 \end{pmatrix}\begin{pmatrix} 0 \\ 3 \end{pmatrix} = \begin{pmatrix} 12 \\ 9 \end{pmatrix}$$

(2-80)

By plotting the points $P'_0 = 0, 0$; $P'_1 = 25, 10$; $P'_2 = 37, 19$; and $P'_3 = 12, 9$; and connecting them by lines, it is seen that we have transformed the rectangle $(P_0P_1P_2P_3)$ into the parallelogram

$(P'_0 \; P'_1 \; P'_2 \; P'_3)$ and the area of the transformed figure is equal to the area of the original figure multiplied by the determinant of the operator. Thus

$$\text{(area) } F_T = \det \begin{pmatrix} 5 & 4 \\ 2 & 3 \end{pmatrix} \text{ (area } F_o) \tag{2-81}$$

where F_T is the transformed figure; F_o is the original figure.

Since in the example, area $F_o = 3 \times 5 = 15$, and $\det \begin{pmatrix} 5 & 4 \\ 2 & 3 \end{pmatrix} =$ $5 \times 3 - 4 \times 2 = 7$, we have area $F_T = (7)(15) = 105$.

All the operations that we have discussed in this section may be extended to apply to 3-dimensional space or even n-dimensional space. For example the operator below may be used to stretch a figure in 3-dimensional space.

$$\begin{pmatrix} k & 0 & 0 \\ 0 & k & 0 \\ 0 & 0 & k \end{pmatrix} \tag{2-82}$$

The equation for this stretching operation is

$$\begin{pmatrix} x_1 \\ y_1 \\ z_1 \end{pmatrix} = \begin{pmatrix} k & 0 & 0 \\ 0 & k & 0 \\ 0 & 0 & k \end{pmatrix} \begin{pmatrix} x_o \\ y_o \\ z_o \end{pmatrix} \tag{2-83}$$

In a similar manner operators may be derived whereby figures or bodies are rotated and reflected in 3-dimensional space, but a detailed account of these operations is beyond the scope of this book. Note, however, that in all cases the determinants of the resulting third order square matrices express the ratio between the volume of the original body and that of the transformed body.

The properties of matrices that we have just discussed provide some indication of the importance of this type of notation for the mathematician, the physicist, and the engineer. In dealing with the problem of stress and strain, for example, matrices can be used to analyze bending problems, e.g., the bending of a bar. Thus the bar may be considered to be composed in section of a large number of small squares, which are transformed into a similar array of parallelograms.

38. Matrices. Other Operations and Definitions. In this section we consider the forms of several types of matrices and then show how an algebra of matrices can be developed to express certain relationships among them.

A square matrix of any order n is said to be *singular* if its determinant is equal to zero and is said to be *non-singular* if its determinant is unequal to zero. Thus the matrix

$$\begin{pmatrix} 6 & 3 \\ 8 & 4 \end{pmatrix}$$

is said to be singular for

$$\det \begin{pmatrix} 6 & 3 \\ 8 & 4 \end{pmatrix} = (24 - 24) = 0$$

and the matrix

$$\begin{pmatrix} 5 & 3 \\ 1 & 2 \end{pmatrix}$$

is said to be non-singular for

$$\det \begin{pmatrix} 5 & 3 \\ 1 & 2 \end{pmatrix} = (10 - 3) = 7$$

We have seen square matrices of the form

$$\begin{pmatrix} A_{11} & 0 & 0 \\ 0 & A_{22} & 0 \\ 0 & 0 & A_{33} \end{pmatrix} \tag{2-84}$$

earlier in this chapter. Such a square matrix of order n, which has non-zero values for the elements along the main diagonal and zero values for all other elements, is defined as a *diagonal matrix* of order n. A special case of this form is called the *unit matrix*. A unit matrix is a diagonal matrix of order n, in which the value of every element on the main diagonal is equal to one, as stated before. For example

$$\begin{pmatrix} 1 & 0 & 0 \\ 0 & 1 & 0 \\ 0 & 0 & 1 \end{pmatrix} \tag{2-85}$$

is a unit matrix of the third order.

The symbol "I" is usually used to represent a unit matrix.

Another example is the transpose of matrix B below

$$B = \begin{pmatrix} m & n \\ o & p \\ q & r \end{pmatrix} \qquad \tilde{B} = \begin{pmatrix} m & o & q \\ n & p & r \end{pmatrix} \qquad \text{(2-99)}$$

Note that while the matrix product AB exists, since the number of columns in A is equal to the number of rows in B, this condition is not true of their transposes, so the product $\tilde{A}\tilde{B}$ does not exist. However, the product $\tilde{B}\tilde{A}$ does exist, as is expressed by the equation

$$F = AB; \quad \tilde{F} = \tilde{B}\tilde{A} \quad \text{(for two matrices and their transposes)}$$

For any number of matrices and their transposes

$$F = ABC\cdots X; \quad \tilde{F} = \tilde{X}\cdots\tilde{C}\tilde{B}\tilde{A}$$

The adjoint of the square matrix A is symbolized by $\hat{A}$, (A), or **adj** A, and is found by replacing every element in the given matrix by its cofactor and then transposing the resultant matrix. For example, if we are given the matrix

$$A = \begin{pmatrix} 4 & 1 & 1 \\ 1 & 2 & 5 \\ 2 & 6 & 4 \end{pmatrix} \text{ then } \hat{A} = \begin{pmatrix} -22 & 2 & 3 \\ 6 & 14 & -19 \\ 2 & -22 & 8 \end{pmatrix} \quad \text{(2-100)}$$

When the relation is expressed symbolically we have

$$A = \begin{pmatrix} a & b & c \\ d & e & f \\ g & h & i \end{pmatrix}$$

$$\hat{A} = \begin{pmatrix} (ei - fh) & -(bi - ch) & (bf - ce) \\ -(di - fg) & (ai - cg) & -(af - cd) \\ (dh - eg) & -(ah - bg) & (ae - bd) \end{pmatrix} \quad \text{(2-101)}$$

It can also be shown that the following relationship holds

$$A\hat{A} = (\det A)\,I$$

To apply this relationship, let us first find $A\hat{A}$ for the example in (2-100)

$$A\hat{A} = \begin{pmatrix} 4 & 1 & 1 \\ 1 & 2 & 5 \\ 2 & 6 & 4 \end{pmatrix} \begin{pmatrix} -22 & 2 & 3 \\ 6 & 14 & -19 \\ 2 & -22 & 7 \end{pmatrix}$$

$$= \begin{pmatrix} -80 & 0 & 0 \\ 0 & -80 & 0 \\ 0 & 0 & -80 \end{pmatrix} \qquad (2\text{-}102)$$

Now we also find that

$$\det A = \det \begin{pmatrix} 4 & 1 & 1 \\ 1 & 2 & 5 \\ 2 & 6 & 4 \end{pmatrix} = -80$$

and $(\det A) I$ is

$$-80 \begin{pmatrix} 1 & 0 & 0 \\ 0 & 1 & 0 \\ 0 & 0 & 1 \end{pmatrix} = \begin{pmatrix} -80 & 0 & 0 \\ 0 & -80 & 0 \\ 0 & 0 & -80 \end{pmatrix} \qquad (2\text{-}103)$$

This relationship leads to another formula that is frequently used for finding the inverse of a square matrix. If A is a square matrix, then as above we have

$$A\hat{A} = (\det A) I$$

dividing through by $(\det A)$ we obtain

$$\frac{A\hat{A}}{(\det A)} = I; \text{ or } A\frac{\hat{A}}{(\det A)} = I$$

We know from the definition of inversion that

$$AA^{-1} = I$$

Thus
$$A^{-1} = \frac{\hat{A}}{(\det A)} \qquad (2\text{-}104)$$

We also have the following definitions.

I — A matrix is said to be *symmetric* if $A = \tilde{A}$

II — A matrix is said to be *skew symmetric* if $A = -\tilde{A}$

III — A matrix is said to be *orthogonal* if $A = \tilde{A}^{-1}$

Note the following examples

$$\begin{pmatrix} 1 & 0 & 1 \\ 0 & 1 & 0 \\ 1 & 0 & 1 \end{pmatrix} \quad \text{is a symmetric matrix}$$

$$\begin{pmatrix} 0 & 1 & 2 \\ -1 & 0 & 1 \\ -2 & -1 & 0 \end{pmatrix} \quad \text{is a skew symmetric matrix}$$

$$\begin{pmatrix} \cos\theta & \sin\theta & 0 \\ -\sin\theta & \cos\theta & 0 \\ 0 & 0 & 1 \end{pmatrix} \quad \text{is an orthogonal matrix}$$

Some of these matrices are used in the subsequent chapters. The primary purpose of this last section has been to show that meaningful relationships can be discovered by treating a matrix as a unit and that an algebra using matrices as the fundamental unit can be successfully developed.

Problems for Solution

1. Add the following pairs of matrices.

a. $\begin{pmatrix} 5 & 7 & 0 \\ 2 & 1 & 6 \end{pmatrix} + \begin{pmatrix} 5 & 0 & 3 \\ 9 & 8 & 10 \end{pmatrix}$

b. $\begin{pmatrix} 1/2 \\ 1/3 \\ 1/4 \end{pmatrix} + \begin{pmatrix} 2 \\ 1/12 \\ 0 \end{pmatrix}$

c. $\begin{pmatrix} -4 & 6 \\ 2.1 & -1 \end{pmatrix} + \begin{pmatrix} 3 & -5 \\ -.9 & 1 \end{pmatrix}$

2. Perform the indicated operation

$$(7)\begin{pmatrix} 6 & 3 & 0 \\ 1 & 5 & 4 \end{pmatrix}$$

3. Multiply the following pairs of matrices:

a. $(2 \quad 4 \quad 6) \times \begin{pmatrix} -5 \\ 3 \\ -1 \end{pmatrix}$

b. $\begin{pmatrix} 4 & 0 & 3 & 2 \\ 5 & 1 & 6 & 1 \\ 3 & 8 & 0 & 2 \end{pmatrix} \times \begin{pmatrix} 1 & 2 \\ 4 & 7 \\ 6 & 5 \\ 3 & 4 \end{pmatrix}$

c. $\begin{pmatrix} -5 & 0 \\ 1/2 & \pi \end{pmatrix} \times \begin{pmatrix} 6 & 4 \\ -1 & -2 \end{pmatrix}$

4. Find the determinants of the following square matrices.

a. $\begin{pmatrix} 5 & 9 \\ 4 & 11 \end{pmatrix}$

b. $\begin{pmatrix} 3a & 2 & 6 \\ 5 & a & -4 \\ 1 & -7 & 3a \end{pmatrix}$

c. $\begin{pmatrix} 1 & 2 & 3 \\ 4 & -1 & 12 \\ 3 & 18 & 9 \end{pmatrix}$

d. $\begin{pmatrix} 2 & 1 & 6 & 3 \\ 1 & 0 & 5 & 2 \\ 3 & 5 & 1 & 1 \\ 2 & 4 & 4 & 5 \end{pmatrix}$

5. Solve by using determinants

a. $\frac{5}{7}x + 6y = 23$

$x - 2/3y = 5$

b. $2x + y = 2.52$

$3x + \frac{1}{11}x = 2.23$

c. $x + 2y + 3z = 29$

$-x + 4y + z = 5$

$2x - y - z = 15$

d. $x + 7y - 2z = 1$

$-2x - 5y + 6z = 4$

$3x - y - 4z = 9$

6. Describe what happens to the size and shape of a rectangle bound by the X-axis, Y-axis, the line $y = 10$, and the line $x = 12$ when it is transformed by the operator

$$\begin{pmatrix} 1 & 2 \\ 1/2 & 1 \end{pmatrix}$$

Also, give the coordinates of the vertices of the transformed figure.

7. Find the inverse of the following matrix by using the formula $A^{-1} = \dfrac{\hat{A}A}{(\det A)}$

$$A = \begin{pmatrix} 2 & 0 & 1 \\ 3 & 1 & 4 \\ 2 & 2 & 5 \end{pmatrix}$$

Chapter 3

PROBABILITY, STATISTICS, AND QUALITY CONTROL

39. Introduction. The subjects of permutations and combinations, and probability are treated in books on arithmetic and algebra, e.g., Thompson, *Algebra for the Practical Man*, 3rd edition, 1962. Consequently, a thorough restatement of the fundamental principles of this subject is not given in the present chapter. However, it is necessary, in the opening sections, to present some concepts which may already be familiar. In discussing these cases theoretical derivations are stated briefly or omitted entirely in order to allow more space for the treatment of new topics.

Probability theory has a wide range of applications and provides a useful analytical tool for science and industry as well as for games of chance. Because of the sweeping scope of this subject, the material presented here is restricted to that which is requisite for an understanding of certain selected applications in the important industrial techniques of quality control, especially that of acceptance sampling.

40. Fundamental Concepts. In this section the essential terminology and prerequisites for the study of probability are summarized and listed in order to provide the reader with the necessary background material for an understanding of the more advanced topics in this chapter.

(A) A set of n distinct numbers, symbols, or events may be arranged in $n!$ different sequences.

Example. The letters A, B, and C may be arranged in the following sequences.

$$ABC \quad ACB \quad BAC \quad BCA \quad CAB \quad CBA$$

$$3! = 3 \times 2 \times 1 = 6$$

Each one of these arrangements is called a *permutation* of the set ABC. A permutation of n things taken all at a time is an ordered arrangement of all the members of the set. The number of possible permutations of n distinct elements, taken n at a time, is equal to $n!$. When some members of the set are alike, the maximum number of permutations of n elements, taken n at a time, is given by the formula

$$\frac{n!}{k_1! \, k_2! \, . \, .}$$

where k_i represents the number of similar elements of each kind.

Example. The elements of the set $AABB$, taken 4 at a time, can be arranged in the following permutations

$$ABAB \quad AABB \quad BABA \quad BBAA \quad ABBA \quad BAAB$$

Here the number of elements in the set, n, is 4, and the number of similar elements is 2 A's and 2 B's, that k_1 and k_2 are both 2. So the formula gives

$$\frac{n!}{k_1! \, k_2!} = \frac{4!}{(2!)(2!)} = \frac{(4 \times 3 \times 2 \times 1)}{(2 \times 1)(2 \times 1)} = \frac{24}{4} = 6$$

(B) A *combination* of a set of objects is any selection of one or more of the objects without regard to order. The number of combinations of n things, taken k at a time, is the number of subsets that can be made up from the set of n things by taking the elements k at a time. Each subset must contain exactly k elements but no two subsets may contain exactly the same elements. The number of combinations of n things taken k at a time is designated by the expression $\binom{n}{k}$ which is equal to

$$\frac{n!}{k!(n-k)!} \tag{3-2}$$

This is an extremely important formula, as becomes evident later in the chapter.

Example. How many combinations can be formed from the set $ABCD$ if the elements are taken 3 at a time?

There are $\binom{4}{3} = \frac{4!}{3! \, 1!}$ combinations

$$= \frac{4 \times 3 \times 2 \times 1}{(3 \times 2 \times 1) \times 1} = 4 \text{ combinations, which are}$$

$$ABC \quad ABD \quad ACD \quad BCD$$

(C) If a given experiment can result in any one of n_1 distinct outcomes and for each of these a subsequent experiment can result in n_2 distinct outcomes, the combined experiment can have $n_1 n_2$ distinct outcomes.

Example. A coin is tossed twice in a row. How many distinct outcomes are there for this combined experiment?

$$2 \times 2 = 4 \text{ distinct outcomes}$$

| Heads-Heads | Heads-Tails |
| Tails-Tails | Tails-Heads |

If each of the outcomes of the first experiment is equally probable and if each of the outcomes of the subsequent experiment is equally probable, then each of the outcomes of the combined experiment is equally probable.

41. Probability. In the different uses and applications of the theory of probability, somewhat different meanings may be attached to the term. For example, if n is the number of mutually exclusive and equally likely outcomes of an experiment, and if m of these outcomes are considered as event A, then the probability of event A occurring is m/n. This fraction m/n, which is defined as the probability of an event, has very little significance for a small number of trials. It indicates rather the frequency with which the event would occur in the course of a very large number of trials. Thus if a perfect coin is tossed 1000 times, it will be found that the ratio of heads to the total number of tosses will very closely approach $\frac{1}{2}$.

However, it is not always possible to count all the ways in which an event might happen. For instance, it is impossible to enumerate all the equally likely ways in which, or reasons why, a person may die at a certain age. It is possible, however, to determine the number of persons out of a large group who die at a certain age, i.e., the number of times the event (death) occurs out of the total possible number. In this case, the ratio of the number of deaths at a certain age, to the total number of people, living and dead, of that age, is the probability of the event (death) at that age. This probability indicates the *frequency* with which events of such a class may reasonably be expected to happen in the case of a very great number of future trials. Thus, if it should be learned from census reports that of each 100,000 persons of age sixty in 1930, about two-thirds were still alive in 1940, it would be said that the probability that a person of age sixty in 1940 would be alive in 1950 is $\frac{2}{3}$.

The fraction m/n is also used to indicate the strength of expectation that an event will occur on a single trial. The greater the ratio of the number of favorable outcomes to the number of possible outcomes, or the greater the frequency with which similar events have happened in the past, the stronger is the expectation that the particular event under consideration will occur on the single trial under consideration.

If we designate p as the probability that an event will happen, then it follows that $(1 - p)$ is the probability that this event will not happen.

Let us consider the throw of a die. The probability that a certain number, say a six, is rolled on a single trial is, of course, $\frac{1}{6}$. Similarly, the probability that an even number is rolled is $\frac{3}{6}$ or $\frac{1}{2}$. This illustrates the important rule, i.e., that if an experiment has a set of n possible outcomes that are mutually exclusive and if m of these outcomes are considered favorable, the probability of obtaining a favorable outcome is equal to the sum of the individual probabilities attached to the m favorable outcomes. Thus, the probability that the throw of a die will result in any of the three numbers, 2 or 4 or 6, is equal to $\frac{1}{6} + \frac{1}{6} + \frac{1}{6} = \frac{3}{6} = \frac{1}{2}$.

Example. Given a bag containing five orange marbles and three black marbles, what is the probability that an orange marble will be selected on a random draw?

There are $5 + 3$ or eight ways in which a marble may be selected and five of these ways will result in the selection of an orange marble. Thus the probability of selecting an orange marble is $\frac{5}{8}$.

Example. Given the same conditions as in the preceding example what is the probability that if two marbles are drawn (a) both are orange (b) one is orange and one is black.

(a) From the preceding section we know that a combination of 2 things may be selected from a set of 8 things in $\binom{8}{2}$ possible ways. Likewise, since we are interested in selecting 2 out of the 5 orange marbles, we know that there are $\binom{5}{2}$ possible combinations that will result in the selection of 2 orange marbles. Therefore the probability of drawing 2 orange marbles is equal to

$$\frac{\binom{5}{2}}{\binom{8}{2}} = \frac{\text{Ways to draw 2 orange marbles out of 5 orange marbles}}{\text{Ways to draw 2 marbles out of 8 marbles}}$$

$$= \frac{5!}{2!(5-2)!} \div \frac{8!}{2!(8-2)!}$$

$$= \frac{5!}{2!\,3!} \times \frac{2!\,6!}{8!}$$

$$= \frac{(5 \times 4 \times 3 \times 2 \times 1)}{(2 \times 1)(3 \times 2 \times 1)} \times \frac{(2 \times 1)(6 \times 5 \times 4 \times 3 \times 2 \times 1)}{(8 \times 7 \times 6 \times 5 \times 4 \times 3 \times 2 \times 1)}$$

$$= \frac{5 \times 4}{2 \times 1} \times \frac{2 \times 1}{8 \times 7}$$

$$= \frac{5 \times 4}{8 \times 7}$$

$$= \frac{20}{56} = \frac{5}{14}$$

(b) To solve this part of the problem we use the same denominator as in part a, i.e. $\binom{8}{2}$ since the total number of possible combinations is the same in each case.   However, we may select one orange marble in $\binom{5}{1}$ different ways and one black marble in $\binom{3}{1}$ different ways.   Thus, the probability of drawing one orange and one black marble is equal to

$$\frac{\binom{5}{1}\binom{3}{1}}{\binom{8}{2}} = \left(\frac{5!}{1!\,4!}\right)\left(\frac{3!}{1!\,2!}\right) \Big/ \left(\frac{8!}{2!\,6!}\right) = \frac{15}{28}$$

Thus, in general, if we define an event (E) as a set of outcomes having a certain property and if the possible outcomes of an experiment are finite in number and all equally likely to occur, the probability that a certain event will occur is given by the formula:

$$P(E) = \frac{\text{Number of possible outcomes in } E}{\text{Total number of possible outcomes}} \qquad (3\text{-}3)$$

Furthermore, if we define $(\sim E)$ as the set of outcomes not in E, then it is obvious that

$$P(E) = 1 - P(\sim E) \tag{3-4}$$

where $P(\sim E)$ is the probability that E will not occur.

Example. If two marbles are drawn from the bag containing five orange and three black marbles, what is the probability that at least one is black?

We may compute this probability most efficiently by employing the formula $P(E) = 1 - P(\sim E)$. In this case $(\sim E)$ is the event that neither of the two marbles is black, or, in other words, it is the event that both of the two marbles are orange. We then have

$$P(\sim E) = \frac{\binom{5}{2}}{\binom{8}{2}} \quad \text{and} \quad P(E) = 1 - \frac{\binom{5}{2}}{\binom{8}{2}}.$$

$$1 - \frac{\binom{5}{2}}{\binom{8}{2}} = 1 - \frac{5}{14} = \frac{9}{14}$$

As we stated earlier in this section, if any one of a mutually exclusive set of outcomes is considered favorable, the probability of obtaining a favorable outcome is $P(E_1 \text{ or } E_2 \text{ or } \cdots \text{ or } E_n) = P(E_1) + P(E_2) + \cdots + P(E_n)$ (where all the outcomes, E_i are mutually exclusive). As an example we said that the probability of rolling an even number with a single die was equal to $P(2 \text{ or } 4 \text{ or } 6) = P(2) + P(4) + P(6) = \frac{1}{6} + \frac{1}{6} + \frac{1}{6} = \frac{1}{2}$.

However, if the criterion for success is based on either of two conditions which are not mutually exclusive, we must employ the following formula.

$$P(E_1 \text{ or } E_2) = P(E_1) + P(E_2) - P(E_1 E_2)$$

This formula may be demonstrated with the aid of Venn diagrams, which were introduced in Chapter 1. For example, if we have two events whose outcomes are mutually exclusive, we have a corresponding Venn diagram as shown in Figure 3-1.

However, if two events occur simultaneously and if the outcomes of the two events are *not* mutually exclusive, we have the situation which is symbolized by the Venn diagram in Figure 3-2.

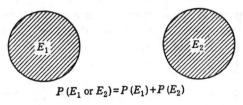

$$P(E_1 \text{ or } E_2) = P(E_1) + P(E_2)$$

Fig. 3-1.

In Figure 3-2 the shaded portion represents those outcomes which occur both in E_1 and in E_2. In the language of set theory, this region would be defined as the intersection of E_1 and E_2. If we consider E_1 and E_2 of Figure 3-2 to be areas and if we add the area of E_1 to the area of E_2, we obtain an area equal to that enclosed by the diagram in

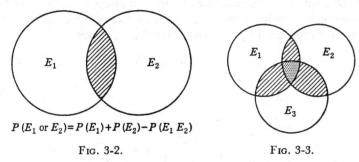

$$P(E_1 \text{ or } E_2) = P(E_1) + P(E_2) - P(E_1 E_2)$$

Fig. 3-2.   Fig. 3-3.

Figure 3-2 plus an additional area of an amount equal to the intersection of E_1 and E_2. We obtain this excess since this common area is a part of both E_1 and E_2 and hence is included twice in the addition process. To correct for this excess, it is necessary to subtract the common term from the sum $E_1 + E_2$.

By extension, if we have three events which are not mutually exclusive, the situation can be symbolized by the Venn diagram in Figure 3-3.

Employing a similar geometrical argument we obtain:

$$P(E_1 \text{ or } E_2 \text{ or } E_3) = P(E_1) + P(E_2) + P(E_3)$$
$$-[P(E_1 E_2) + P(E_1 E_3) + P(E_2 E_3) - P(E_1 E_2 E_3)]$$

or

$$P(E_1 \text{ or } E_2 \text{ or } E_3) = P(E_1) + P(E_2) + P(E_3)$$
$$- P(E_1 E_2) - P(E_1 E_3) - P(E_2 E_3) + P(E_1 E_2 E_3)$$

Thus, in general, it follows that, for n simultaneous events which are *not* mutually exclusive, the probability that the outcome of a given experiment will satisfy the criteria for *at least one* of these events is given by the formula:

$$P(E_1 \text{ or } E_2 \text{ or } E_3 \text{ or } \cdots \text{ or } E_n) = \sum_{i=1}^{n} P(E_i) - \sum_{\substack{i=1 \\ i \neq j}}^{n} P(E_iE_j) + \cdots$$
$$\pm P(E_1E_2E_3 \cdots E_n) \quad (3\text{-}5)$$

Example. Consider the set of all integers between and including one and ten. If one of these integers is selected at random, what is the probability that it will be even *or* divisible by three?

Let A be the set of all integers between and including one and ten and let E_1 be a subset of A containing all the even integers in A and let E_2 be a subset of A containing all the integers in A divisible by three. This situation is illustrated by the Venn diagram in Figure 3-4.

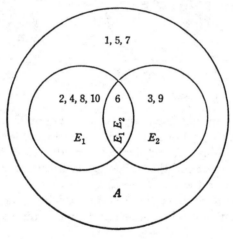

FIG. 3-4.

Thus we have $P(E_1 \text{ or } E_2) = P(E_1) + P(E_2) - P(E_1E_2)$

$$P(E_1 \text{ or } E_2) = P(even) + P(divisible\ by\ 3)$$
$$- P(even\ and\ divisible\ by\ 3)$$

$$P(E_1 \text{ or } E_2) = \frac{1}{2} + \frac{3}{10} - \frac{1}{10} = \frac{7}{10}$$

Another useful device for computing simple probabilities is provided by the multiplication law.   Let us again consider the bag containing 5 orange and 3 black marbles.   If two marbles are drawn, what is the probability that both are orange?

Previously we found that the solution to this problem was equal to $\binom{5}{2} \Big/ \binom{8}{2} = \frac{5}{15}$. However there is another avenue of approach available to us.   We can easily see that the probability of drawing an orange marble on the first attempt is $\frac{5}{8}$.   Then if we do succeed in drawing an orange marble on the first attempt, the probability of drawing a second orange marble is $\frac{4}{7}$.   To determine the combined probability, simply multiply these two figures together.

$$5/8 \times 4/7 = 5/14$$

Thus, in general, we have the formula:

$$P(E_1 \text{ and } E_2) = P(E_1)P(E_2 \mid E_1) \tag{3-6}$$

where $P(E_2 \mid E_1)$ is read as "E_2 given E_1."

Many problems in probability cannot be categorized easily and solved by the basic formulae.   However, there is one type of problem which occurs frequently and which can be solved by following a special rule.   This rule applies provided that we desire to know the probability that an event (E_1) will occur in a specified number of a finite number of trials.   The outcome for each trial is independent of all previous outcomes and the probability that E_1 will occur in a single trial is known and must be constant for each trial.   If a given situation satisfies these criteria, it is then possible to apply the *binomial formula*, which states that the probability that E_1 will occur k times in n trials equals:

$$P(E_{\binom{n}{k}}) = \binom{n}{k}p^k(1 - p)^{n-k} \tag{3-7}$$

where p is the probability of E occurring in a single trial.

Example.   What is the probability that if 3 coins are tossed, at most two heads show?

Three results will satisfy the above condition: 1) no heads show 2) one head shows 3) two heads show.   Thus, if we designate these possibilities E_1, E_2, and E_3 respectively, we have

$$P(E_1 \text{ or } E_2 \text{ or } E_3) = P(E_1) + P(E_2) + P(E_3)$$

$$P(E_1) = \binom{3}{0}\left(\frac{1}{2}\right)^0\left(\frac{1}{2}\right)^3 = \frac{3!}{0!(3-0)!}(1)\left(\frac{1}{8}\right) = \frac{1}{8}$$

$$P(E_2) = \binom{3}{1}\left(\frac{1}{2}\right)^1\left(\frac{1}{2}\right)^2 = \frac{3!}{1!(3-1)!}\left(\frac{1}{2}\right)\left(\frac{1}{4}\right) = \frac{3}{8}$$

$$P(E_3) = \binom{3}{2}\left(\frac{1}{2}\right)^1\left(\frac{1}{2}\right)^1 = \frac{3!}{2!(3-2)!}\left(\frac{1}{4}\right)\left(\frac{1}{2}\right) = \frac{3}{8}$$

$$P(E_1) + P(E_2) + P(E_3) = \frac{1}{8} + \frac{3}{8} + \frac{3}{8} = \frac{7}{8}$$

42. Games of Chance. No discussion of probability can be said to be complete without a brief account of its origins. The fact is that the early problems from which the modern theory of probability developed were questions asked by gamblers, usually of eminent mathematicians of the time.

One of the most famous was the problem of the unfinished game which was presented to Blaise Pascal. Pascal corresponded about it with Fermat, and this correspondence established the basic method of attack upon questions in probability.

Players A and B play games (the nature of the game is immaterial; it can even be matching pennies) for a total stake of $120, which is to be won by the first player to win 3 games. After the score has reached 2 for A and 1 for B, they are called away. What is the fair division of the $120 between them?

The original distribution proposed to Pascal was 2 to 1, that is $80 to A and $40 to B, since A had won 2 games to B's 1 game. In proving that this division was incorrect, Pascal and Fermat arrived at the procedure of evaluating the possibilities, and thus established the basis for any probability analysis.

Suppose that the contest is continued to completion. With a score of A, $2 - B$, 1, if A wins the next game he will obtain $120. If the game is entirely one of chance (i.e., the skill of the players is not a factor), A has a 1 : 2 chance of winning this fourth game and the $120. Obviously this chance is worth $(1 : 2)(\$120) = \$60$ to A.

Suppose that A loses the fourth game, of which there is also a 1 : 2 chance. Then he still has a 1 : 2 chance of winning the fifth game and the $120. Thus his chance of losing the fourth game and winning the fifth is $(1 : 2)(1 : 2) = (1 : 4)$, and the value of this chance is $(1 : 4)(\$120) = \$30$.

Therefore A should take the value of these two chances, which is $60 + $30 = 90, leaving only $30 for B. Computation of the value of B's only chance of success, that is winning two games in succession, would obviously yield the same amount, $(1 : 2)(1 : 2)(\$120) = \$30$.

One of the most famous problems in probability was proposed by Nicolaus Bernoulli, and developed at length by Daniel Bernoulli in the Transactions of the St. Petersburg Academy, whence it has always been called the St. Petersburg paradox.

A coin is tossed until heads appears. If heads appears on the first toss, the bank pays the player $1. If heads appears for the first time on the second toss, the bank pays $2. If heads appears for the first time on the third toss, $4; on the fourth toss, $8; on the fifth toss, $16; and so on. What amount should the player pay the bank for the privilege of playing one game in order that the game be fair — that is to say, in order that neither the player nor the bank has an advantage regardless of how long the game goes on?

Consider the first toss of the coin. The probability of heads is $\frac{1}{2}$. The amount involved is $1. Therefore the expectation on this toss is $\frac{1}{2}$ of $1, or $\frac{1}{2}$ dollar. Consider the second toss. The player will collect on this toss only if he throws tails on the first toss and heads on the second. The probability that this will happen is $(\frac{1}{2})(\frac{1}{2})$, or $\frac{1}{4}$. The amount involved is $2. Therefore the expectation on this toss is $\frac{1}{4}$ of $2, or $\frac{1}{2}$ dollar. Consider the third toss. The player will collect on this toss only if he throws tails on the first two tosses and heads on the third. The probability that this will happen is $(\frac{1}{2})(\frac{1}{2})(\frac{1}{2})$, or $\frac{1}{8}$. The amount involved is $4. Therefore the expectation on this toss is $\frac{1}{8}$ of $4, or $\frac{1}{2}$ dollar.

To show that the expectation on every toss is $\frac{1}{2}$ dollar, consider the nth toss. The player will collect on this toss only if he throws tails on the first $n - 1$ tosses and heads on the nth. The probability that this event will happen is $(\frac{1}{2})(\frac{1}{2})(\frac{1}{2}) \cdots (\frac{1}{2})$ to n factors, or $(\frac{1}{2})^n$. Now the number of dollars involved in the first toss is 1, or 2^0; that in the second toss, 2, or 2^1; that in the third toss, 4, or 2^2; that in the fourth toss, 8, or 2^3; and so on. Note that the number of dollars is always a power of 2, and that the power is always one less than the number of the toss

Hence the number of dollars involved in the nth toss is 2^{n-1}. Finally, then, the expectation on the nth toss is $(\frac{1}{2})^n (2^{n-1})$, or $2^{n-1}/2^n$, or $\frac{1}{2}$ dollar.

Since the total expectation is always the sum of the expectations at each stage of the game, the total expectation here is

$$\tfrac{1}{2} + \tfrac{1}{2} + \tfrac{1}{2} + \tfrac{1}{2} + \tfrac{1}{2} + \tfrac{1}{2} + \cdots \text{ dollars.}$$

Now recall that play is to continue until heads turns up. Theoretically there is no limit to the number of tails which may appear before the first head appears, and this means that the above series is to be summed to infinity. But the sum of the infinite number of terms of this series is obviously infinite. It follows that the player must pay the bank an infinite amount of money for the privilege of playing one game!

Obviously this result is absurd, but the reason for its absurdity is far less obvious. In fact, a number of approaches have been suggested. One of the simplest is to point out that we have dealt with an infinite number of $\$\frac{1}{2}$ terms as the possible winnings, although no bank can have an infinite amount of money. If we take a large sum, say $\$1,000,000$, as the total resources of the bank, then the problem yields an acceptable solution.

We can arrive at this solution by writing an expression for the series of expectations, using p_i to represent the probability of a given payment and s_i to represent the amount of the payment, so that the expectations are the product terms $p_i s_i$. The probabilities are inverse powers of 2, and the payments are powers of 2 (taking $\$2^0 = \$1$ as the first payment.) Also $\$2^{19}$ is less than $\$1,000,000$ ($\$524,288$) and $\$2^{20}$ is greater than $\$1,000,000$ ($\$1,048,576$) so that the terms after the twentieth have payments of $\$1,000,000$.

The total expectation is therefore

$$\tfrac{1}{2}(\$2^0) + \frac{1}{2^2}(\$2) + \frac{1}{2^3}(\$2^2) + \ldots + \frac{1}{2^{20}}(\$2^{19})$$

$$+ \frac{1}{2^{21}}(\$1,000,000) + \frac{1}{2^{22}}(\$1,000,000) \cdot \ldots$$

Multiplying out the first twenty terms, we obtain

$$(\$\tfrac{1}{2}) + (\$\tfrac{1}{2}) + (\$\tfrac{1}{2}) \cdots (\$\tfrac{1}{2})$$

Since this part of the series contains twenty terms, each $\$\frac{1}{2}$, its sum is $\$10$.

Therefore the total expectation simplifies to

$$\$10 + \frac{1}{2^{21}}(\$1,000,000) + \frac{1}{2^{22}}(\$1,000,000) \ldots .$$

Moreover, the series of terms after the $10 constitute an infinite geometric progression, the formula for the sum of which is

$$S = \frac{a}{1 - r}$$

where a is the first term and r is the ratio.
Substituting we have

$$S = \frac{\frac{1}{2^{21}}(\$1,000,000)}{1 - \frac{1}{2}}$$

$$= \frac{1}{2^{20}}(\$1,000,000)$$

$$= \frac{\$1,000,000}{1,048,576} = \$.954$$

Therefore, the expectation is $10 + $0.954 = $10.95, which represents the sum a player should pay for the right to play to a million dollar bank if neither is to have an advantage.

43. Random Walk Processes. Random walk processes are significant, not only in many types of physical phenomena, but also in basic statistical methods, including decision making and the related sequential sampling. The name "random walk" is derived from a simple physical description of the theory.

Consider that O is a point on a straight line, that A is a point on the line a paces to the right of O, and that A' is a point on the line a paces to the left of O. Then if one walks on the line by taking one pace at a time, each pace being taken at random to the right or the left, how many paces will be required to reach either A or A', which terminates the walk?

Obviously if the number of paces from O to A, and that from O to A' are great, the number of paces before the walk terminates will also be great, since because the paces are randomly directed toward A or toward A', many a pace will retrace the ground covered by the previous pace. However, the probability of oscillating to and fro without ever reaching A or A' can be shown to be zero.

A formulation of the random walk function in terms of time, which is more generally useful in physical processes, may be made as follows. If a person makes a step of length h every r seconds and each step is equally likely to be to the right or to the left, then the probability at time t of being at a distance of x from where he was at time $t = 0$ can be shown to be given by a function $U(x, t + r)$ which satisfies the difference equation

$$U(x, t + r) = \tfrac{1}{2}U(x + h, t) + \tfrac{1}{2}U(x - h, t). \qquad (3\text{-}8)$$

The function U can be approximately evaluated by letting a computing machine "make" a large number of random walks by reference to a sequence of random numbers. (See page 116 for a discussion of random numbers).

A random walk is a *stochastic process*, which is a general term for a chain of events which change in a random way with time. Sequential binomial sampling is another example of a stochastic* process, since each "step" in the "walk" represents the result of sampling one more item from the population to determine whether it is to be accepted or rejected. The decision to be reached by the process is to determine whether or not to accept a certain hypothesis about the population. Such a hypothesis might relate to the proportion of defective members, the proportion of members having a property within specified limits, or some similar statement.

44. Markov Processes. A Markov process is any stochastic process in which the future development is completely determined by the present state and not at all by the way in which the present state arose. A Markov chain is a sequence of Markov processes. Consider, for example, the "tree" of events formed by tossing a coin three times, and shown in Figure 3-5. Any pathway through the tree from left to right constitutes a Markov chain.

The definition may be expressed mathematically by stating that the value of a variable x at any time t_i depends at most on its value at the preceding time t_{i-1}. (Both time intervals and values of the variable

---

*For practically all ordinary purposes, the word "stochastic" may be replaced by the word "chance."

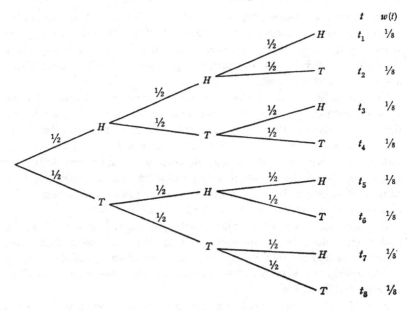

t	$w(t)$
t_1	$\frac{1}{8}$
t_2	$\frac{1}{8}$
t_3	$\frac{1}{8}$
t_4	$\frac{1}{8}$
t_5	$\frac{1}{8}$
t_6	$\frac{1}{8}$
t_7	$\frac{1}{8}$
t_8	$\frac{1}{8}$

Fig. 3-5.

x are taken to be discrete.)   The joint probability for an observed set of values x_1, x_2, $\cdots$ x_n, is thus

$$P = P(x_1) \cdot P(x_2 \mid x_1) \cdot P(x_3 \mid x_2) \cdots P(x_n \mid x_{n-1})$$

so
$$P = P(x_1) \prod_{r=2}^{n} P(x_i \mid x_{i-1}) \tag{3-9}$$

where the last equation states that the probability of the entire Markov chain (of discrete events) is the product of the separate probabilities of the individual events that constitute the chain.

45. Random Variables.   A *variate* or *random variable* is a variable whose value can not be predicted.   However, a random variable must have a designated range of possible values with a known probability corresponding to each value.   There are numerous examples of random variables.   For instance, in a given population, the members differ in height, weight, reading speed, age, income, etc.   Each of these characteristics is a random variable which assumes values in an unpredictable fashion if the given population is sampled at random.

There are two distinct types of random variables. A *discrete random variable* can only assume integer values. Falling into this category are size of family, number of pages in a telephone book, the spots appearing on a die that is rolled, or, in short, any phenomenon whose possibilities are finitely denumerable.

On the other hand, a *continuous random variable* is a characteristic which is measurable and which can take on all possible values within limits on a given scale of measurement. Examples are the height of adult males, the amount of time an individual spends driving an automobile during a given period, the volume of water used daily by a given town, etc. In reality it is impossible to obtain a perfectly continuous distribution because of the limitations of any measuring device and because the sample being studied may be too small.

Random variables are usually denoted by capital letters (X, Y, Z), and their values by small letters. Thus the expression $P(X = a)$, where a is a real number which has been assigned to a possible outcome of an experiment, is read as "the probability that the outcome is a" or "the probability of a."

Example. Consider drawing a card at random from a typical deck containing 52 cards. The following numerical values are then assigned to the possible outcomes.

Card drawn	Value
club	1
diamond	2
heart	3
spade	4

In this case, the random variable X can assume any one of these four values and

$$P(X = 1) = P(\text{club}) = \frac{1}{4}$$

$$P(X = 2) = P(\text{diamond}) = \frac{1}{4}$$

$$P(X = 2 \text{ or } 3) = P(\text{diamond or heart}) = \frac{1}{4} + \frac{1}{4} = \frac{1}{2}$$

$$P(X = 1 \text{ or } 3 \text{ or } 4) = P(\text{club or heart or spade}) = \frac{1}{4} + \frac{1}{4} + \frac{1}{4} = \frac{3}{4}$$

The distribution function of a random variable describes the cumulative frequency with which a set of outcomes occurs and may be graphed by utilizing the equation

$$F(x) = P(X \leq x) \tag{3-10}$$

which reads "the cumulative frequency of x is equal to the probability that the value of the random variable X is less than or equal to x."

Example. Let us again consider the single draw of a card from a deck of 52 cards. The distribution function of the random variable corresponding to the four possible outcomes described above, appears in Figure 3-6.

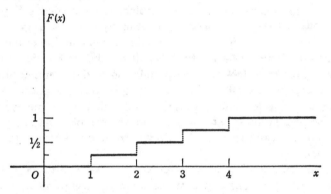

Fig. 3-6.   The distribution function of a discrete random variable.

This graph is based on the equation

$$F(x) = P(X \leq x) = \begin{cases} 0 \text{ if } x < 1 \\ \dfrac{1}{4} \text{ if } x < 2 \text{ and } x \geqslant 1 \\ \dfrac{1}{2} \text{ if } x < 3 \text{ and } x \geqslant 2 \\ \dfrac{3}{4} \text{ if } x < 4 \text{ and } x \geqslant 3 \\ 1 \text{ if } x \geqslant 4 \end{cases}$$

The stepping stone pattern of this graph is typical of a discrete distribution function, i.e. the distribution function of a discrete random variable. In contrast, the distribution function of a continuous random variable describes a continuous curve such as shown below in Figure 3-7.

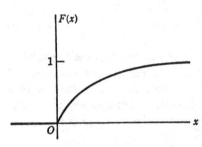

This curve satisfies the following continuous distribution function.

$$F(x) = 0 \quad \text{if} \quad x \leq 0$$

$$F(x) = 1 - e^{-x} \quad \text{if} \quad x \geq 0 \quad *$$

46. Statistical Definitions and Their Notation. The term population means the entire collection with which a particular statistical problem is concerned. It may be indefinitely great, or as great as the inhabitants of a nation or the world,

Fig. 3-7. The distribution function of a continuous random variable.

or it may be limited to a relatively small number of entities; thus the fifty-two cards in a deck was the population of the example discussed in the preceding section. The term sample means a portion chosen from the population, usually with the implication that some hypothesis about the population is to be drawn from the sample, or some hypothesis about the sample is to be drawn from our knowledge of the population.

Suppose that a discrete variate X can take k distinct values x_i (where $i = 1, 2, \cdots, k$) and that there are f_i individuals in a sample having the value x_i. The total size of the sample is $\sum f_i = N$. Then the rth *moment of X about zero* is defined as

$$m_r' = \frac{1}{N} \sum_{i=1}^{k} f_i x_i^r \qquad (3\text{-}11)$$

An important case is where $r = 1$, whence the expression simplifies to

$$m_1' (\text{often written } m) = \frac{1}{N} \sum_{i=1}^{k} f_i x_i \qquad (3\text{-}12)$$

which is merely the arithmetic mean. (Note that the symbol m, usually written for the arithmetic mean of a population, is commonly replaced by $\bar{x}$ for the arithmetic mean of a sample.

*The letter "e" represents the base of natural logarithms a transcendental number which when carried out to 10 decimal places, is equal to 2.7182818285. (See Chapter 1)

If X (the random variate) is discrete and if the probability is p_j that it takes the value x_j, then the rth *moment of X about* 0 is defined as

$$\mu_r' = \sum_j x_j^r p_j = E(X^r) \tag{3-13}$$

where $E(X^r)$ is the expectation of X^r. (In other words, this summation gives both the rth moment of X about zero and the expectation value.)

Here again the first moment is the expectation of X and is commonly denoted simply by μ, just as the first moment of X in equation (3-12) was denoted by m. The practice of using Greek letters for population measures and English letters for sample statistics is common in statistics.

If X is a continuous random variable (instead of a discrete one) equation (3-13) becomes

$$\mu_r' = E(X') = \int_a^b x^r f(x)\, dx \tag{3-14}$$

where a and b define the range of the probability density function of (x). (See section on normal distribution for further discussion of continuous random variables and the density function.) Another statistical measure is the moment about the mean. The rth *moment of X about the mean* is defined as

$$m_r = \frac{1}{N} \sum_{i=1}^k f_i (x_i - m)^r \tag{3-15}$$

Here the first moment

$$m_1 = \frac{1}{N} \sum_{i=1}^k f_i (x_i - m) \tag{3-16}$$

is of little value as a statistical measure, since it must equal 0 for every sample, since $f_i x_i = m$ by definition of the arithmetic mean. However, the second moment of X about the mean is an important measure called the variance

$$m_2 = \frac{1}{N} \sum_{i=0}^k f_i (x_i - m)^2$$

$$= \frac{1}{N} \left(\sum_{i=0}^k f_i x_i^2 - 2m \sum_{i=0}^k f_i x_i + m^2 \sum_{i=0}^k f_i \right)$$

$$= m_2' - m^2 \tag{3-17}$$

where m'_2 is the second moment of a sample about zero and m is the arithmetic mean. (Note that the variance of a population is denoted by μ_2).

The positive square root of the variance is the standard deviation, denoted by s for a sample and σ for a population. Thus the variance may also be found by squaring the standard deviation, and hence written as s^2 or σ^2. For the mean or expectation of the values x_i of the variate X over a population, the symbol μ_x is used, while the standard deviation is commonly expressed by σ_x^2.

47. Important Distributions. The Binomial Distribution. Earlier in this chapter we introduced the binomial formula (Equation (3-7)) which states that the probability of an event E occurring k times in n trials is equal to

$$P(X = k) = \binom{n}{k} p^k (1 - p)^{n-k} \qquad (3\text{-}18)$$

Here p is the probability of occurrence of an event or characteristic on a single trial and $(1 - p) = q$ is the possibility of its not occurring. A random variable having this type of probability function is said to have binomial distribution. If we assign the value 1 to X if E does happen and the value 0 to X if it does not and if we have $P(X = 1) = p$ and $P(X = 0) = (1 - p) = q$ for any single try, we obtain

$$E(X) = \mu_x = np \qquad (3\text{-}19)$$

where n is the total number of tries. Likewise for the variance and standard deviation, we obtain

$$\sigma_x^2 = npq \qquad (3\text{-}20)$$

$$\sigma_x = \sqrt{npq} \qquad (3\text{-}21)$$

Example. If there is a 2% probability that a part produced by a given machine is defective, what is the probability that none out of a sample of six will have a defect? Also for a sample of 100, what is the expected number of defects, and the corresponding variance and standard deviation?

First we assign the value 1 to the outcome that a selected part has a defect and the value 0 to the outcome that a selected part does not have a defect.   Then the probability of no defects in a sample of six by Equation (3-19) is equal to

$$P(X = 0) = \binom{6}{0}\left(\frac{1}{50}\right)^0\left(\frac{49}{50}\right)^6 = \left(\frac{49}{50}\right)^6 = .885$$

For a random sample of 100 parts we have

$$E(X) = \text{(expected number of defects)} = np = \frac{100}{50} = 2$$

$$\sigma_x^2 = (100)(1/50)(49/50) = \frac{49}{25}$$

$$\sigma_x = 7/5$$

The Hypergeometric Distribution.   There are two basically different types of sampling processes.   In the first type an article is removed from the lot, tested, and replaced before another article is drawn.   The number in the sample having a certain property E is then found to be binomially distributed.   However, if a group of articles is selected simultaneously, i.e. there is no replacement between tests, the type of distribution is then found to be *hypergeometric*.   The probability of there being k articles having the characteristic E in a sample of size n is then given by the equation

$$P(X = k) = \frac{\binom{M}{k}\binom{N - M}{n - k}}{\binom{N}{n}} \tag{3-22}$$

where N is the lot size of which M have the property E.   For this type of distribution the expected value has the same form as for the binomial distribution

$$E(X) = \mu_x = n\left(\frac{M}{N}\right) \tag{3-23}$$

The variance and consequently the standard deviation differ, however, from the form used when calculating for a binomial distribution. Here we have

$$\sigma_x^2 = \left[n\left(\frac{M}{N}\right)\right]\left(\frac{N - n}{N - 1}\right)\left(\frac{N - M}{N}\right) \tag{3-24}$$

$$\sigma_x = \sqrt{n\left(\frac{M}{N}\right)\left(\frac{N - n}{N - 1}\right)\left(\frac{N - M}{N}\right)} \tag{3-25}$$

Example. If we have a sample of 8 parts from a lot of 20, 4 of which are known to be defective, (a) what is the probability that 2 of the parts in the sample are defective and (b) what is μ_x, σ_x^2, and σ_x for the sample of 8.

a) Applying the probability equation (3-22) we obtain

$$P(X = 2) = \frac{\binom{4}{2}\binom{16}{6}}{\binom{20}{8}} = .381$$

b) And by Equations (3-23), (3-24) and (3-25), we obtain

$$\mu_x = 8\frac{4}{20} = \frac{8}{5} = 1.6$$

$$\sigma_x^2 = \left[\frac{8}{5}\right]\left(\frac{12}{19}\right)\left(\frac{16}{20}\right) = \frac{384}{475} = .808$$

$$\sigma_x = \sqrt{\frac{384}{475}} = .899$$

The hypergeometric distribution, as well as the binomial distribution, is based upon a discrete random variable and therefore is characterized as a discrete distribution.

The Poisson Distribution. Another important discrete frequency distribution used in statistical practice is the Poisson distribution, a special case of the binomial distribution where p is small and n is large. It can be used to describe such phenomena as the distribution of bacterial colonies by very small unit areas in a culture spread over a large number of such areas, the distribution of the emission of particles per unit of time of a radioactive substance over a considerable number of such units of time, and the distribution of telephone calls received per minute at a switchboard over a period of several hours.

The Poisson distribution is derived from the binomial distribution by assuming that $p = \dfrac{\bar{x}}{n} = \dfrac{\mu_x}{n}$ where μ_x is constant, p is small and n is large. If we substitute $p = \dfrac{\mu_x}{n}$ in the expression for the binomial distribution:

$$\binom{n}{k}p^k(1 - p)^{n-k}$$

we obtain

$$\frac{n!}{k!(n-k)!}\left(\frac{\mu_x}{n}\right)^k\left(1-\frac{\mu_x}{n}\right)^{n-k}$$

Rearranging the terms gives

$$\frac{n!}{n^k(n-k)!}\left(\frac{\mu_x{}^k}{k!}\right)\left(1-\frac{\mu_x}{n}\right)^n\left(1-\frac{\mu_x}{n}\right)^{-k} \tag{3-26}$$

As n, the size of the random sample goes to infinity, the first and last terms approach 1 while the third term approaches $e^{-\mu_x}$, so that the probability of any value of x is

$$e^{-\mu_x}\frac{\mu_x{}^k}{k!} \tag{3-27}$$

This formula is usually written as

$$P(X = k) = e^{-\lambda t}\frac{(\lambda t)^k}{k!} = p(k \text{ changes in interval of width } t) \tag{3-28}$$

where λ is a proportionality constant characteristic of the given problem. The sum of all the probabilities in a Poisson distribution is, of course, equal to one. To prove this, we recall the definition of e^x.

$$e^x = \sum_{k=0}^{\infty}\frac{x^k}{k!}$$

Thus, when $x = \lambda t$, we have

$$e^{\lambda t} = \sum_{k=0}^{\infty}\frac{(\lambda t)^k}{k!} \tag{3-29}$$

Then the sum of all the probabilities in the given distribution is

$$\sum_{k=0}^{\infty}p(k) = \sum_{k=0}^{\infty}e^{-\lambda t}\frac{(\lambda t)^k}{k!} = e^{-\lambda t}\sum_{k=0}^{\infty}\frac{(\lambda t)^k}{k!}$$

$$= e^{-\lambda t}e^{\lambda t} = 1 \tag{3-30}$$

For any given situation the product (λt) is a constant regardless of the units in which t is expressed. Thus, (λt) is often taken as the average number of times an event occurs in an interval of length t. It may also be taken as the expected number of failures in a sampling process. If we set $t = 1$, then $\lambda =$ mean number of times an event occurs in a unit interval.

Example. A piece of equipment is to be exposed to operating conditions for a period of 300 hours with a .05 failure rate and a corresponding MTBF (Mean Time Between Failures) of 20 hours. What is the probability that, over any randomly selected period of 10 hours, there will be no failures?

In this case, the variable t is time and the random variable X is the number of failures in 10 hours. Since we know that the average (MTBF) is one failure in 20 hours, we have

$$\lambda t = E(X) = \mu_x = \frac{1}{2} \text{ failure in 10 hours}$$

The value of λt remains constant and is independent of the units selected to represent t. However λ does vary according to the units selected for t. If one hour is the unit adopted, λ is equal to the average number of failures in one hour, i.e. .05. We then have

$$t = 10 \qquad \lambda = .05 \qquad \lambda t = .5 = 1/2$$

If, on the other hand, we had selected 50 hours to be the unit of time, then a period of 10 hours would be designated as $\frac{1}{5}$ and λ would be the average number of failures in 50 hours, i.e. $\frac{5}{2}$. Thus we would have

$$t = 1/5 \qquad \lambda = 5/2 \qquad \lambda t = 1/2$$

We have now established that λt is a constant equal to $\frac{1}{2}$ for this problem.

By Equation (3-28), we obtain:

$$X = 0,$$
$$P(\text{no failures in 10 hours}) = P(X = 0)$$
$$= e^{-1/2} \frac{(1/2)^0}{0!} = e^{-1/2} = .606531$$

If the problem had been to find the probability of there being *at most* one failure in the period of 10 hours, we would have had

$$P(X \leq 1) = P(X = 0) + P(X = 1)$$
$$= e^{-1/2} \left[\frac{(1/2)^0}{0!} + \frac{(1/2)^1}{1!} \right]$$
$$= e^{-1/2} \left[1 + \frac{1}{2} \right]$$
$$= \frac{3}{2} e^{-1/2} = .909795$$

One very interesting feature of the Poisson distribution is that, for any given problem, both the expected value and the variance have the same value, i.e. λt. Consequently, the standard deviation is equal to $\sqrt{\lambda t}$.

A great advantage of the Poisson notation is that it can be used to approximate a binomial distribution where n is very large and p is very small. If this criterion is satisfied, the following approximation may be made.

$$\binom{n}{k} p^k (1 - p)^{n-k} \doteq e^{-np} \frac{(np)^k}{k!} * \tag{3-31}$$

Example. If a machine turns out parts with a failure rate of .01, what is the probability that in a sample of 500, at most 2 parts are defective?

If we let the random variable X represent the number of parts which are defective out of the sample of 500, we have a binomial distribution and the corresponding probability is given by

$$P(X \le 2) = P(X = 0) + P(X = 1) + P(X = 2)$$

$$P(X = 0) = \binom{500}{0} (.01)^0 (.99)^{500}$$

$$P(X = 1) = \binom{500}{1} (.01)^1 (.99)^{499}$$

$$P(X = 2) = \binom{500}{2} (.01)^2 (.99)^{498}$$

$$P(X \le 2) = \binom{500}{0} (.01)^0 (.99)^{500} + \binom{500}{1} (.01)^1 (.99)^{499}$$
$$+ \binom{500}{2} (.01)^2 (.99)^{498}$$

$$\tag{3-32}$$

In many problems, the amount of computation needed to evaluate such expressions is very great. Consequently it is obviously more convenient to settle for a close approximation provided by the Poisson formula.

*The symbol $\doteq$ represents the expression, "is approximately equal to."

Thus we have:

$$P(X \leq 2) \doteq P(X = 0) + P(X = 1) + P(X = 2)$$

$$np = 500(.01) = 5$$

$$P(X \leq 2) \doteq e^{-5}\left[\frac{(5)^0}{0!} + \frac{(5)^1}{1!} + \frac{(5)^2}{2!}\right]$$

$$\doteq e^{-5}\left[1 + 5 + 12\frac{1}{2}\right]$$

$$\doteq e^{-5}[18.5] \doteq .007638(18.5)$$

$$\doteq .1246530$$

(3-33)

Because of the availability of tables of the exponential and the Poisson functions, this method has become one of the most widely used in the calculation of probability or frequency distributions for the purposes of industry and science.

Furthermore, the hypergeometric distribution can be approximated by the binomial distribution. When the N in the hypergeometric formula becomes very large, the fraction M/N, which remains constant, can be represented by the p in the binomial formula provided that n is small compared to N. Consequently, the procedure for finding the desired probability can be further simplified, without a significant increase in error, by using the Poisson formula to approximate the binomial formula if the binomial expression satisfies the criterion stated above.

Example. If a bin contains 10,000 screws of which 3% are defective, what is the probability that there will be exactly 5 defects in a sample of 100 screws which are selected at random but without replacement?

This problem has the form of a hypergeometric distribution and as such can be represented by Equation (3-20)

$$P(X = 5) = \frac{\binom{300}{5}\binom{9700}{95}}{\binom{10000}{100}}$$

where N = Lot size = 10,000

M = Number of defects in lot = $.03(N)$ = 300

n = Sample size = 100

k = Number of defects in sample = 5

The following criteria are satisfied

1) N is very large

2) N is large in comparison with n

3) M/N is constant

In order to approximate the hypergeometric solution, we let $M/N = p$ and express the problem in terms of the binomial Equation (3-18)

$$P(X = 5) \doteq \binom{100}{5} (.03)^5 (.97)^{95}$$

n = Sample size = 100

k = Number of defects in sample = 5

p = Proportion of defects in lot = .03

We can go a step further and find $P(X = 5)$ by the Poisson equation, since the conditions satisfy the criterion for a Poisson approximation, i.e. n is very large compared to p.   Thus we let $\lambda t = np = 3$ and obtain:

$$P(X = 5) \doteq e^{-3} \frac{(3)^5}{5!} = .049787\left(\frac{243}{120}\right)$$

$$\doteq .100819 \quad \text{(Poisson)}$$

The Normal Distribution.   All of the distributions discussed so far have been of the discrete type, i.e. based on a discrete random variable.   In this section we discuss a continuous frequency distribution.   The distribution function of a continuous random variable is continuous and differentiable at all but a finite number of points.   The derivative of the distribution function is referred to as the *density function* and measures the concentration of probability at a given point.   If we let $F(x)$ be the cumulative distribution function and if we let $f(x)$ be the density function of $F(x)$, we have

$$\int_a^b f(x)dx = F(b) - F(a) \tag{3-34}$$

since we know that $f(x) = \dfrac{dF(x)}{dx}$.

However, the expression $F(b) - F(a)$ may be interpreted as the probability that x will assume a value in the interval a, b. Thus we have

$$\int_a^b f(x)dx = F(b) - F(a) = P(a \leqslant x \leqslant b) \qquad (3\text{-}35)$$

The most important of the continuous distributions is the normal distribution whose density function is given by the formula:

$$f(x) = \frac{1}{\sigma_x \sqrt{2\pi}} e^{-1/2(x-\mu_x/\sigma_x)^2} \qquad (3\text{-}36)$$

In this equation σ_x and μ_x are parameters. A *parameter* is defined as an arbitrary constant in a mathematical expression, which distinguishes various specific cases by assuming different values. Thus, in the equation for a straight line, $y = mx + b$, m and b are parameters which specify particular straight lines.

If in the equation for the density function of normal distribution we select $\sigma_x = 1$ and $\mu_x = 0$ as parameter values we obtain the function

$$f_\phi(x) = \frac{1}{\sqrt{2\pi}} e^{-x^2/2} \qquad (3\text{-}37)$$

The probability distribution corresponding to this density function is called *standard normal distribution*. In defining the parameters we have already specified that the expectation of this distribution (μ_x) is 0 and that the standard deviation (σ_x) is 1. Consequently the variance (σ_x^2) is also 1.

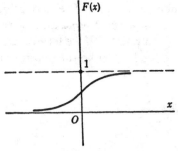

FIG. 3-8a. Standard normal distribution function.

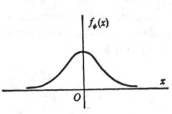

FIG. 3-8b. Density function for standard normal distribution.

The general form of the standard normal distribution function is shown in Figure 3-8a. Note that the mean or expectation of x is 0. The general form of the density function corresponding to standard normal distribution is shown in Figure 3-8b. Note that this function and hence the probability is also centered around zero.

Given any continuous random variable which has a density function in the normal form, it is possible to "standardize" this variable by subtracting its mean and dividing this difference by its standard deviation. This introduces the new random variable Z.

$$Z = \frac{X - \mu_x}{\sigma_x} \tag{3-38}$$

The function based on this variable fits the criteria for standard normal distribution, i.e. $\mu_z = 0$ and $\sigma_z = 1$. Hence, as a result of the introduction of this variable, the center of the distribution is over the origin and the basic unit on the horizontal axis becomes one standard deviation.

However, whether or not a normal distribution is standard, the area under the curve formed by its density function is always equal to one. This area corresponds to the total probability of the function. Thus to find the probability that the value of the random variable Z lies between a and b, we simply take the area under the density curve between these two values. It is important to remember that, for the density curve of a standard normal distribution, half of the total area lies on each side of the line $Z = 0$ (see Figure 3-9).

In Table 3-1 the values of the ordinates and areas of the standard normal curve are given. In the first column we have z, which represents the values that the random variable Z might take. In the second column $f_\phi(z)$ gives the corresponding ordinate (perpendicular distance from the point on the horizontal axis to the density curve).

In the third column $\int_0^z f_\phi(z)dz$ gives the area under the curve between $Z = 0$ and $Z = z$. In Figure 3-9, the shaded area under the curve represents the probability that $0 \leq z \leq +1$ and is given by the corresponding figure in the third column of Table 3-1. The ordinate corresponding to $z = 1$ is the value

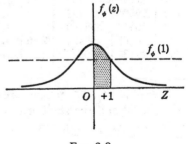

Fig. 3-9.

on the curve, vertically above the point (+1) on the Z-axis. This value is given in the second column of Table 3-1.

TABLE 3-1

Ordinates and Areas of the Normal Curve

$$f_\phi(z) = \frac{1}{\sqrt{2\pi}}\, e^{-z^2/2}$$

z	$f_\phi(z)$	$\int_0^z f_\phi(z)\,dz$	z	$f_\phi(z)$	$\int_0^z f_\phi(z)\,dz$	z	$f_\phi(z)$	$\int_0^z f_\phi(z)\,dz$
.00	.39894	.00000	1.35	.16038	.41149	2.70	.01042	.49653
.05	.39844	.01994	1.40	.14973	.41924	2.75	.00909	.49702
.10	.39695	.03983	1.45	.13943	.42647	2.80	.00792	.49744
.15	.39448	.05962	1.50	.12952	.43319	2.85	.00687	.49781
.20	.39104	.07926	1.55	.12001	.43943	2.90	.00595	.49813
.25	.38667	.09871	1.60	.11092	.44520	2.95	.00541	.49841
.30	.38139	.11791	1.65	.10226	.45053	3.00	.00443	.59865
.35	.37524	.13683	1.70	.09405	.45543	3.05	.00381	.49886
.40	.36827	.15542	1.75	.08628	.45994	3.10	.00327	.49903
.45	.36053	.17364	1.80	.07895	.46407	3.15	.00279	.49918
.50	.35207	.19146	1.85	.07206	.46784	3.20	.00238	.49931
.55	.34294	.20884	1.90	.06562	.47128	3.25	.00203	.49942
.60	.33322	.22575	1.95	.05959	.47441	3.30	.00172	.49952
.65	.32297	.24215	2.00	.05399	.47725	3.35	.00146	.49960
.70	.31225	.25804	2.05	.04879	.47982	3.40	.00123	.49966
.75	.30114	.27337	2.10	.04398	.48214	3.45	.00104	.49972
.80	.28969	.28814	2.15	.03955	.48422	3.50	.00087	.49977
.85	.27798	.30234	2.20	.03547	.48610	3.55	.00073	.49981
.90	.26609	.31594	2.25	.03174	.48778	3.60	.00061	.49984
.95	.25406	.32894	2.30	.02833	.48928	3.65	.00051	.49987
1.00	.24197	.34134	2.35	.02522	.49061	3.70	.00042	.49989
1.05	.22988	.35314	2.40	.02239	.49180	3.75	.00035	.49991
1.10	.21785	.36433	2.45	.01984	.49286	3.80	.00019	.49993
1.15	.20594	.37493	2.50	.01753	.49379	3.85	.00024	.49994
1.20	.19419	.38493	2.55	.01545	.49461	3.90	.00020	.49995
1.25	.18265	.39435	2.60	.01358	.49534	3.95	.00016	.49996
1.30	.17137	.40320	2.65	.01191	.49598	4.00	.00013	.49997

Example. If 3000 people are given an examination for which the mean is 50 and the standard deviation is 10, how many grades 75 or over can be expected, assuming normal distribution?

The formula for standardizing a normal distribution is, from Equation (3-38)

$$Z = \frac{X - \mu_x}{\sigma_x}$$

Thus we have:

$$Z = \frac{X - 50}{10}$$

But X is given as 75, so

$$Z = \frac{75 - 50}{10} = \frac{25}{10} = 2.5$$

The percent likely to score 75 or over equals the probability of scoring 75 or better and this is equal to the area under the normal curve between $Z = 2.5$ and $Z = +\infty$. This, in turn is equal to the expression

$$P(Z \geq 75) = \frac{1}{2} - \int_0^{2.5} f_\phi(z)dz$$

In this expression the area under the normal curve between $Z = 0$ and $Z = 2.5$ is subtracted from one-half of the total area under the curve, which as stated above, is 1. Thus, using Table 3-1, to evaluate the integral, we find for 2.50 in the first column, .49379 in the third, so

$$P(Z \geq 75) = \frac{1}{2} - .49379$$

$$= .50000 - .49379$$

$$= .00621 = .621\% \text{ (or about 19 people)}$$

Example. For the same examination, what per cent can be expected to score 30 or less?
Again we have:

$$Z = \frac{X - 50}{10}$$

But $X = 30$

$$Z = \frac{30 - 50}{10} = -2$$

Since the normal curve is symmetrical with respect to the line $Z = 0$, we may consider the solution to this problem to be represented by the area under the curve between $Z = 2$ and $Z = +\infty$. Thus we have:

$$P(Z \le 30) = \int_2^{\infty} f_\phi(z)dz = \frac{1}{2} - \int_0^2 f_\phi(z)dz$$

$$= \frac{1}{2} - 47725$$

$$= .50000 - .47725$$

$$= .02275 = 2.275\% \text{ (or about 68 people)}$$

Example. For the same examination what is the probability of scoring exactly 50?

We cannot solve this directly for the probability of hitting one point out of an infinite number is zero. Similarly the area under a point is equal to zero. Thus we must consider the score of 50 to be that score lying in an interval. Let us choose the interval between a score of 49.5 and one of 50.5. Since 50 is the mean in this problem the area under the curve between 50 and 50.5 is equal to the area under the curve between 49.5 and 50. Thus we find one of these two areas and double the result to obtain the solution. Thus we have:

$$Z = \frac{50.5 - 50}{10} = \frac{.5}{10} = .05$$

$$P(50 \le Z \le 50.5) = \int_0^{.05} f_\phi(z)dz = .01994 \text{ (which is the value in}$$
third column of table for .05 in the first column)

$$P(Z = 50) = P(49.5 \le Z \le 50.5) = 2(.01994) = .03988$$

Thus 3.998% of the people, or about 120, can be expected to score between 49.5 and 50.5.

48. Elements of Statistics. *Statistics* is that branch of mathematics which deals with the accumulation and analysis of quantitative data. There are three principal subdivisions in the field of statistics but these overlap, more often than not, in actual practice. First, inference from samples to population by means of probability is called *statistical inference*. Second, *descriptive statistics* is defined as the characterization and summarization of a given set of data without direct reference to inference. And finally, *sampling statistics* deals with methods of obtaining samples for statistical inference. It is this last subdivision that we will consider first in this section.

Sampling and the Use of Random Numbers. As used in statistics, the word *sample* is a part of the (entire) statistical population which is selected at random and is used for making estimates and inferences about the population. One of the most important branches of statistics deals with the principles of sampling which can be applied to obtain sound estimates of the various types of populations.

For the populations that are of interest in statistics vary widely in character and attributes. A relatively simple type of population is a finite group, however large, of denumerable objects, such as the residents of a community or the articles produced during a day (or week or month) by a machine that fabricates parts or assemblies of only one kind. However, many statistical populations are classed as infinite, meaning that they are not conveniently denumerable or are indefinitely large, or both, as the population of molecules in a body of gas (or liquid or solid) or the number of times that a coin or die can be thrown. Some populations are fixed or static (at least for the time under consideration), as the residents of a community or the oil in a tank, while others are changing or dynamic, as the customers in a store or the oil in a pipe line. Some populations are known, insofar as the frequency distribution of the statistical variable under investigation is concerned; others are unknown, and in fact, most statistical problems arise in an effort to estimate these unknown distributions by means of random sampling.

Before discussing random sampling there are two points to be made about sampling in general. Sampling is the selection of units from a population. Now the division into units may be effected by the sampling process itself, for while a population of persons in a city or cards in a pack already consists of discrete units, the sampling unit for oil or coal is the container used in taking the samples.

The second point is that most real sampling, as that of oil or coal, is sampling "without replacement." This is different from many statistical experiments, as made with cards, colored balls, etc., in which the population is kept constant by replacing the element which is drawn before proceeding with the next draw.

The definition of random sampling or randomization is "a process in which each sampling unit in the population has an equal chance of being selected at every draw." This process can be effected, theoretically at least, in certain simple cases by mechanical means only, as in the oft-cited experiments with cards and colored balls. In dealing with more extensive and complex populations, other means are necessary. In the case of a population consisting of a finite number of objects which can be counted and numbered, such as persons, houses, machines, the first step is to obtain or prepare a list of all the individual elements, and then to number them consecutively, as is shown in the annual incomes of the members of a club given in Table 3-2.

TABLE 3-2

Annual Incomes of Membership of XYZ Club

1 Allen	8 Dudley	15 Johnson
$6200	$6800	$8000
2 Altherton	9 Eisenhauer	16 Martin
$9400	$7600	$9000
3 Billings	10 Felici	17 Nilsson
$4600	$6800	$4900
4 Brooks	11 Frank	18 Roberts
$8000	$9000	$8800
5 Carter	12 Gilbert	19 Simpson
$10,000	$6000	$6900
6 Cerano	13 Grant	20 Sternfels
$4200	$7200	$5000
7 Drake	14 Hastings	21 Thomson
$7200	$5300	$10,000

Now suppose it is required to pick a random group of 6 members for the purpose of estimating the average income, that is, the arithmetic mean income, of the entire population. The procedure is to apply a table of random members in such a way as to fit the numbers 1-21 of the club members. Tables of random numbers are published in various statistical publications in various arrangements. Table 3-3 is reprinted from "Elementary Principles of Statistics," by Rosander, (Van Nostrand).

To use this table to choose a random sample of 6 of the 21 club members, run down the table, column by column, and note the first six of the numbers from 1-21 which appear. They are 3, 14, 19, 21, 8, and 4, and by applying these random numbers to the sequentially

TABLE 3-3

Non-repeating random numbers
001–200

41	185	107	1	13	52	151	115	134	194
167	73	81	68	120	146	70	64	31	118
125	147	21	183	27	172	193	72	76	65
197	162	140	93	144	98	139	37	131	33
181	36	8	123	106	178	90	30	35	200
133	159	4	165	160	117	53	198	92	116
3	108	32	85	7	135	173	86	42	132
49	58	169	109	124	5	82	62	168	95
112	129	96	78	67	84	138	128	195	15
149	113	148	142	45	186	10	155	40	34
14	110	69	54	164	163	174	24	189	6
66	48	12	25	190	60	114	75	158	71
199	19	136	44	170	157	122	121	111	18
126	192	103	184	127	99	153	187	9	191
87	104	188	150	38	171	56	154	89	101
137	97	51	88	166	91	28	17	23	29
50	161	63	119	74	55	105	175	57	182
180	39	130	77	83	61	11	16	176	177
59	47	80	152	22	196	79	179	46	43
94	102	143	20	26	145	2	100	141	156

numbered list of club members, we choose Billings, Hastings, Simpson Thomson, Dudley, and Brooks. Their incomes are, respectively, $4600; $5300; $6900; $10,000; $6800 and $8000. These six figures have an arithmetic mean of $6933, to the nearest dollar, against an arithmetic mean for the entire population of 21 members of $7105, which is a close approximation in view of the wide variation in the figures of the population. For that reason, one would not, of course, use such a sample merely to reduce the work of computation from 21 to 6 numbers. However, in many statistical problems the figures can not be determined for an entire population, and then some sampling method is essential.

In choosing random samples from larger populations, slight modifications of this method are often necessary. Thus to choose a sample from the telephone listings of a great city, we would need to use a list of random numbers three times, once for pages, a second time for columns and a third time for lines in the columns. Moreover, we would need to use a special device, such as the first or last random number digit only for columns, to avoid the time spent in finding the small digits representing the columns in a large list of random numbers. For in this case the list would need to be much larger than the 001-200 in our table, since the method is only fair if all the position numbers of the population appear in the random list. For the number of pages in a large-city telephone book, this would require a random number list well into the thousands.

Various methods have been used for the construction of tables of random numbers. Tippetts selected at random from census reports 40,000 digits and combined them by groups of 4 into 10,000 numbers. Fisher and Yates in their *Statistical Tables* (Oliver and Boyd) gave pairs of random numbers, tabulated in groups of five pairs, obtained from the right side figures (15-19th places) in a twenty-place logarithmic table. Kendall and Babington Smith used a special machine to construct 100 groups of 1000 digits each. In this case, however, their check of randomness disclosed that five of their 100 groups were not satisfactory when used by themselves.

This test of randomness is an important criterion, although it is not entirely sufficient by itself. To apply this test count the number of repetitions of each digit (including 0) and determine if it occurs in normal expectation (which is $\frac{1}{10}$, since there are ten digits, 0–9.). Consider, for example, the twenty five-digit numbers below:

91624	42761	53819	70052	97369
24917	53813	89703	45631	79240
40118	55026	12859	68347	60995
71448	84900	36075	27263	25808

By counting the frequency of occurrence of each digit in this list we obtain the following frequency distribution

Digit:	0	1	2	3	4	5	6	7	8	9
Frequency:	11	10	10	9	10	10	9	10	10	11

Since there are 100 digits in the 20 five-digit numbers, by normal expectation each of the ten digits would occur $\frac{100}{10} = 10$ times. It is apparent from the above frequency distribution that the departure from normal expectation is sufficiently small for a group of this size, so that this group of numbers has passed this test of randomness. Further tests can be made by counting the frequency of pairs of the same digit (in the same number and in consecutive numbers) as well as in other groupings.

The method of randomization by the use of random numbers is relatively easy to apply, as shown above, when the population consists of discrete entities that can be numbered. With other populations special methods must be devised. For grain in a bin or oil in a tank, one such method is to divide the volume of the filled portion of the bin or tank geometrically into cells of equal size, and to use a special sampler to withdraw a sample from each. Then we can assign consecutive sample numbers to each such sample, and select any required number of them by the use of a random table. An analogous situation exists in agricultural experimentation in which we wish to test different seeds or fertilizers or insecticides in different plots on a field, and wish to devise a sampling system that will yield test results that are not colored by the chance variables of the field, such as variations of soil quality, moisture, etc., from plot to plot. Here we use the method of Latin Squares, that is, we divide the field into square plots of equal size, number them consecutively, and use a table of random numbers to select as many plots as we wish to study.

The basis of all these methods of sampling non-discrete populations by tables of random number tables is the use of a geometrical (or other mathematical) device to divide the population into discrete parts.

49. Other Fundamental Concepts. As has already been indicated, the complete set of observations upon which a statistical analysis is based is usually called a "sample of n," where n refers to the number of observations. The sample of observations is usually assumed to be representative of a much larger number of possible observations or measurements that might be made under the same experimental conditions. This larger group of potential observations is called a "population." Measurements both of the population and of the sample are distributed in some way from a minimum value to a maximum value. Since a statistical distribution has a central tendency and a spread, evaluations of these in the form of measures of central value and dispersion.

Among the measures of central value are (1) the *arithmetic mean* or *average* which was described earlier in this chapter; (2) the *mode*, or that observation which in the distribution has the highest frequency; and (3) the *median*, or the observation which has 50% of the other observations below it and 50% above it.

The two most important measures of dispersion are (1) the sample standard deviation

$$\text{deviation } s = \sigma_s = \sqrt{\frac{\sum (x_i - \bar{x})^2}{n - 1}} \text{ and (2), the range } R = (x_{max} - x_{min})$$

(3-39)

Another important measurement is *skewness*, which provides an indication of the symmetry of a distribution. A distribution is symmetric only if its mean, median, and mode all coincide. Distributions in which the mode is less than the median show "positive" skewness, while, if the situation is reversed, the distribution is said to show negative skewness. The most common method of measuring skewness is to take the ratio of the difference, mean minus mode, to the standard deviation

$$\text{skewness} = \frac{\text{mean} - \text{mode}}{\text{standard deviation}}$$

(3-40)

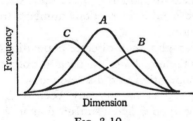

Fig. 3-10.

In Figure 3-10 left, A designates the graph of a normal distribution of values of one dimension of a number of samples of a machine part. In this case mean = median = mode and skewness = 0. Similarly, B designates the graph of a distribution which is negatively skewed and C designates the graph of a distribution which is positively skewed.

50. Statistical Quality Control. *Statistical Quality Control* is a method by which the quality of a manufactured product may be controlled. Its main purpose is to trace and eliminate systematic variations, the remaining variation then being of the nature of a distribution of error; and the properties of the latter can be measured and used to guarantee the average quality of the product.

In order to trace the causes of defectiveness in the product and to keep a constant check on its quality, several types of control charts are used by the quality control engineer.

The "R" chart measures the range and hence the uniformity or consistancy of a product. The narrower the R chart, the more uniform is the product. Considered to be the most sensitive control graph, the R chart is the best method of detecting erratic conditions and general statistical instability.   Among the situations that will have an effect upon an R chart are (1), a new, poorly trained or careless operator or inspector; (2), a faulty machine; (3), faulty material (4), faulty testing equipment.

To construct an R chart, it is first necessary to select the sample size (n).   As a rule the sample size should not exceed ten units.   Then, a series of samples are obtained.   Preferably there should be at least 15 samples before a graph is constructed.   For each sample R is computed by subtracting the lowest measurement from the highest measurement. The average of all the R's are taken and this value $\bar{R}$ becomes the center line of the R chart.   Then the desired control limits are drawn above and below $\bar{R}$.   Figure 3-11 is an example of an R chart.   Here sample 9 is out of control, i.e., its range is not within the bounds imposed by the control limits.

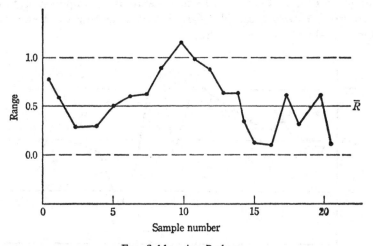

Fig. 3-11.   An R chart.

Another important graph is the $\bar{x}$ chart (x-bar chart).   This is a measurement of the average value per sample of one characteristic under observation.   The $\bar{x}$ chart indicates where a certain operation or process is centered.   If the $\bar{x}$ chart is normal the various sample values cluster around the center line.   However, values which are gradually moving up or down, or which are erratic indicate loss of control. The $\bar{x}$ chart can be affected by such factors as (1) change of operator or inspector; (2) change in material; (3) change in machine setting; (4) change in atmospheric or other environmental conditions; or (5) wear of tool. The $\bar{x}$ charts should only be evaluated after first taking into consideration the readings of the R chart, since some abnormalities which are evident on the R chart cause apparent changes on the $\bar{x}$ chart.

To construct the $\bar{x}$ chart, the same samples and sample sizes as those used in the R chart are employed again. This time the value of $\bar{x}$ is found for each sample and the average of all the $\bar{x}$'s, $(\bar{\bar{x}})$, is the center line of the $\bar{x}$ chart. Finally the desired control limits are drawn in, using broken lines, and the points are plotted and connected with straight line segments. It is usually a good plan to use similar scales in constructing the R and $\bar{x}$ charts, for they are often used in combination to control a process.

Figure 3-12 is an $\bar{x}$ chart. The first 5 and last 5 samples seem to be in control. However samples 6-10 indicate an upward trend and samples 11-15 are somewhat erratic.

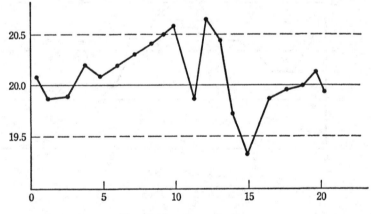

Fig. 3-12.   An $\bar{x}$ chart.

The third important quality control graph is the p chart, which usually measures the percent defective in a sample. The p chart can be affected by such variations as (1) fluctuations in the material used; (2), a faulty machine part or fixture; (3), change in methods or routine; (4), change in machine operator. A change on a p-chart may indicate either that the percentage of defective products is increasing or decreasing, or that there has been a change in the criteria for an acceptable product.

To construct a p-chart it is necessary to use the following values.

Sample size, denoted by n

Fraction defective in sample, denoted by p

Average fraction defective in a series of samples, denoted by $\bar{p}$.

A useful p-chart must record the p of at least 10 samples of uniform size. Whereas the sample size used in constructing the R and $\bar{x}$ charts were small, those used in constructing the p-chart should be considerably larger, e.g., 50 or 100. The number of defective units in each sample is counted and the value of p is computed and recorded. The average value of p, $(\bar{p})$, is calculated and becomes the center line of the p-chart. The upper and lower control limits are then calculated from the following formulas

$$UCL = \bar{p} + 3\sqrt{\frac{\bar{p}(1 - \bar{p})}{n}} \tag{3-41}$$

$$LCL = \bar{p} - 3\sqrt{\frac{\bar{p}(1 - \bar{p})}{n}} \tag{3-42}$$

An appropriate scale is then selected and the points are plotted and connected by straight line segments. An example of a p-chart is given in Figure 3-13. This section has treated some of the fundamental techniques used in the detection of systematic variations of a product. In the next and concluding section of this chapter another important

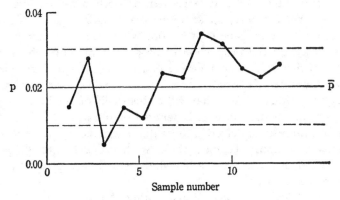

FIG. 3-13. p-chart

aspect of statistical quality control, i.e., acceptance sampling, is discussed.

Acceptance Sampling. One of the major industrial control activities is the decision, on the basis of a random sample, whether or not a given lot is of acceptable quality. In a factory acceptance sampling has two specific applications. First, small parts manufactured by other concerns may be tested to ensure that they meet the required quality specifications. Second, small parts, semi-assembled products, and the finished product manufactured at the plant itself may be tested to ensure a continued level of quality from the plants own machines, and inspectors.

If a large quantity of a given product were 3% defective, we might expect to find an average of 3 defective units in each sample of 100. However, we would not expect to find exactly three defective units in each sample of 100 because of the non-uniform distribution of the units. Thus, the quality control engineer might decide to accept any lot provided that a sample of 100 revealed 4 or fewer defective units. As we shall see if the lots are consistently 3% defective and the *acceptance number* for a sample of 100 is 4, approximately 80% of the lots presented will be accepted. However, if the lots become 7% defective and 4 is maintained as the acceptance number, only about 17% of the lots presented will be accepted.

Thus when a higher percentage of defects begins to appear in lots of a certain product, the sampling process, provided that the same acceptance number is maintained, will not usually reject all of the product submitted but only a greater proportion as the quality diminishes. Furthermore, the sampling plan does not guarantee that a lot which is accepted is better than one which is rejected since such a plan is based on the proportion of product, represented by a set of samples, which will, over a period of time, be accepted or rejected.

The sample size is usually denoted, as usual, by the symbol "n". This is the number of units that will be inspected out of a lot of a given size. The acceptance number is denoted by the symbol "c" and is the maximum number of defective units that a sample can have if the lot is to pass inspection. Together these two figures define the proportion of a product of a certain quality which will be accepted or rejected over a period of time.

The *probability of acceptance* of a sampling plan is the proportion of a product accepted under the given conditions of sample size, acceptance number, and percent defective.   Usually denoted by the symbol Pa and expressed as a decimal, the probability of acceptance for a product of any quality can be calculated on the basis of any combination of sample size and acceptance number.

The equation most frequently used to calculate the probability of acceptance is the Poisson formula.   An excellent source of numerical Poisson probabilities is the book *Tables of the Individual and Cumulative Terms of Poisson Distribution* by Members of the Staff General Electric Co. (D. Van Nostrand Company, Princeton, N. J. 1962).

Recall the Poisson Equation (3-26)

$$P(X = k) = e^{-\lambda t} \frac{(\lambda t)^k}{k!}$$

If $\lambda t = np = U =$ expected number of defectives in a sample of size n then,

$$P(X = k) = e^{-\lambda t} \frac{(\lambda t)^k}{k!} = e^{-np} \frac{(np)^k}{k!} = e^{-U} \frac{(U)^k}{k!} \tag{41}$$

The "k" in this expression corresponds to the acceptance number of c.

Thus the probability of acceptance for a sample plan in which the expected number of defectives in a sample of size n is pn or U and the acceptance number is c or k is given by the formula

$$Pa(X \leq k) = e^{-U} \left(\frac{(U)^1}{0!} + \frac{(U)^1}{1!} + \dots + \frac{(U)^k}{k!} \right) \tag{42}$$

In the book of tables suggested above the cumulative probability as well as individual probabilities are tabulated, which this greatly simplifies the solution of most problems.

Example.   Consider the sampling plan $n = 100$, $c = 6$.   What is the probability of acceptance when the product is 9% defective?

$$U = np = 100 \times .09 = 9$$

$$c = k = 6$$

$$Pa(X \leq 6) = e^{-9} \left[\frac{9^0}{0!} + \frac{9^1}{1!} + \frac{9^2}{2!} + \frac{9^3}{3!} + \frac{9^4}{4!} + \frac{9^5}{5!} + \frac{9^6}{6!} \right]$$

If we use the Poisson tables, we simply look up $U = 9$, avoiding this computation.   Then we look down the left hand column to $x = 6$ and across to the corresponding reading under the column headed $C(x)$. This column gives the cumulative probability from 0 to x for the given

U. Thus we find $C(6) = .20678$. This then is the desired probability and it may be interpreted to mean that for an incoming product that is 9% defective with samples of size 100 being inspected with an acceptance number of 6, only about 20% of the product will be accepted over a period of time. To calculate the probability of rejection it is of course only necessary to subtract the probability of acceptance from unity.

The probability of rejection is usually referred to as the "Producer's Risk." This probability represents the risk of rejecting a product when the lot quality is comparatively good.

Example. Consider the sampling plan, $n = 100$, $c = 4$. What is the Producer's Risk (i.e. probability of rejection) if the units are normally 2% defective?

Using the tables ($U = 2$; $x = 4$) we find that the probability of acceptance is .94735. Thus the probability of rejection is $1 - .94735 = .05265$ and the Producer's Risk is 5.265%. The engineer attempts to minimize the producers risk so that there will be less danger of rejecting lots that are good.

Similarly, for every sampling plan, there is a "Consumer's Risk" which is defined as the probability of accepting the product when the lot quality is poor. The engineer also attempts to minimize the Consumer's Risk of a product.

Example. Using the same sampling plan as in the previous example, i.e., $n = 100$, $c = 4$, what is the Consumer's Risk if the consumer desires to reject a product which is 8% defective?

This is, of course, the probability of acceptance where $U = 8$ and $x = 4$. The solution is .09963 or about 10%. However this figure does not indicate that the consumer has a 10% chance of receiving a defective product. It does mean that a product (8% defective) submitted to inspection according to the above sampling plan would have only a 10% chance of being accepted.

When the quality of a product being manufactured is consistently good, it is wise for the engineer to adopt a sampling plan which will minimize the Producer's Risk. If a good product is rejected too often the results can be costly and time-consuming. Among the possible effects of this situation are

    I. Loss of Time
        a. rechecking
        b. reloading + unloading

     c. Interruption of routine

     d. Procuring needed parts

  II. Loss of Space

 III. Loss of quality because of repeated handling

However, if the quality of the product being manufactured is statistically abnormal or inconsistent, the emphasis should be placed on minimizing the Consumer's Risk. Thus in many cases, there will be two possible sampling plans for a given product, one to use when production is running smoothly and one to use when it is erratic.

Problems for Solution

1. If a box contains 7 good and 2 bad batteries, what is the probability of drawing 3 good batteries in succession?

2. For the same situation as given in problem 1, what is the probability of drawing at least 2 good batteries if 3 are selected at random?

3. Let us say that a product can have any combination of 3 defects (i.e. it may have none, 1, 2, or 3 of the cited defects). If we know that in a lot of 100 units

> 9 units have defect A
> 16 units have defect B
> 11 units have defect C
> 6 units have defect A and defect B
> 5 units have defect A and defect C
> 9 units have defect B and defect C
> 4 units have all three defects

What is the probability that an article drawn at random will have one or more defects?

Hint — Construct a Venn Diagram to represent the problem.

4. (a) What is the probability of rolling a pair of 3's with 2 dice?

   (b) What is the probability of rolling at least one 3 when 2 dice are rolled simultaneously?

5. What is the probability that a coin will come down heads at least 3 times in 5 flips?

6. Describe the distribution function of a random variable which can assume the values 0 and 1 with equal probability (i.e. $\frac{1}{2}$).

7. A die is rolled 2880 times. What is the mean and standard deviation of the number of three's thrown.

8. If a bin contains a large number of assemblies that are known to be 10% defective, what is the probability that there would be at most one defect in a sample of 3?

9. If from a lot of 14 parts, of which 3 are defective, a sample of 6 is taken, what is the probability that there will be exactly one defective unit in the sample?

10. A retailer with limited storage space finds that, on the average, he sells two boxes of parrot food per week. He replenishes his stock every Monday morning so as to start the week with four boxes on hand. What are the probabilities that (a) he sells his entire stock in a week (b) he is unable to fill at least one order? (c) With how many boxes should he start the week so as to have a probability of at least 0.99 of being able to fill all orders. Hint: assume a Poisson distribution of sales with mean 2, and find the probability of x or more sales.

11. If on the average the proportion of defective fuses in a large consignment is 0.015, calculate the approximate probability that in a box of 200 fuses there will not be more than 2 defective.

12. The mean height of soldiers in a regiment containing 1000 men is 68.22 in., with a standard deviation of 3.29 in. If the distribution is normal, how many men over 6 ft. tall would you expect to find in the regiment?

13. If a company wishes to accept no more than 27% of a product which is 5% defective, what acceptance number should it use for samples of 100? Using the acceptance number, what is the probability of acceptance if the product is only 2% defective?

Chapter 4

THE THEORY OF GAMES

51. Introduction. The fundamental importance of the theory of games is clearly apparent from the statement that this discipline made possible, for the first time, a mathematics of competition and a basis for making sound strategic decisions in competitive situations. Such situations arise in all phases of human activity: in business, economics, war — in short, in any situation involving the interplay of two or more policies.

The theory of games was presented comprehensively in the book, "Theory of Games And Economic Behavior", written in 1944 by John von Neumann and Oskar Morgenstern, although the subject was treated in earlier mathematical papers by those authors and others, and it has since been extended in many later publications. Its methods are essentially mathematical in character; in fact, the objective of the book by von Neumann and Morgenstern was to "mathematize economics" in the sense of providing mathematical methods for the formulation and, as far as possible, the solution of certain economic problems. In fact, the authors considered those problems as being reducible to essentially a single problem, that of the "maximization of utility" or in the case of business, of the "maximization of profit."

To appreciate the importance of this problem to economics, consider the effect of successive increments in the number of businesses producing and/or marketing a particular product. When this number (in a closed economy or isolated market) is 1, the situation is that described by the word *monopoly* as used in classical economics, and its optimum course of action is determined from the principles and logic of classical economics. In mathematical terms, the analysis of an isolated business may be made in terms of certain physical data such as: resources, raw materials, demand, etc., and the problem is to apply this data so as to obtain a maximum result. While, of course, even a monopoly cannot maximize interdependent variables, such as volume of production, gross income, and profits, simultaneously, it can choose one of these, such as profits, and adjust the others to maximize the one. This is an ordinary maximum problem, in which the difficulties, however great, are purely technical.

Now, if the number of businesses making or selling the product in the market is increased by 1, so that there is a "duopoly" instead of a monopoly, the situation becomes entirely changed. Here are present, not only the variables of each business, but those arising from the relations between them. That is, since their actions affect each other, then the result for each one depends not merely upon its own actions, but also upon those of the other. Thus, each business endeavors to maximize a function of which it does not control all the variables. This, therefore, is no maximum problem, but a mixture of such problems, and there is no basis for its solution in classical economics.

When a third economic participant, e.g., a third business, is added to the two, the situation becomes still more complex, since there are now three sets of partial variables concerned. Furthermore, with each additional competitor that is added, the complexity continues to grow, until the number becomes so great that the influence of each becomes negligible, and the principles of statistics and probability are applicable, at least to the extent to which they have been elaborated for economic problems. This final condition would realize, of course, the free competition of classical economics. There is left, therefore, as the province of a new field of investigation, the gap between the monopoly (single enterprise economy) on one hand and the N-enterprise economy on the other.

This problem to which von Neumann and Morgenstern addressed themselves is of very wide occurrence, in business as in other fields. Their approach utilized various mathematical procedures, notably those of set theory, functional analysis, and mathematical logic. However, they and many of the later contributors to the subject have furnished many explanations and examples, whereby there is available a basis for a non-analytical survey of the methods and results.

The basic approach is familiar from having been used before in exploring new fields or in attacking old unsolved problems. It is to simplify, to the utmost extent possible, the problem to be solved, while retaining its essential features. Since the distinguishing characteristic of this problem is that of competition, that characteristic can be understood by the study of competitive games. While it is true that games do not approach in complexity the many intricate elements of an economic or business problem, for that very reason they provide a means of attacking mathematically any such problem of conflicting interests and strategies.

As a further step in simplifying the analysis, the theory classifies games into two major groups: *zero-sum games* and *non-zero-sum games*. In a zero-sum game the sum of the payments between the players as of the end of the game is zero, that is, one player's gain is a loss to the other or the others. All games played for entertainment belong to this class, while competitive situations in business rarely do. The development of the theory of zero-sum games was found to provide methods of analysis which could be extended to the non-zero-sum ones.

Several fundamental questions arise in any discussion of game theory. These include the following: How does each player plan his strategy? How much information is available to each player throughout the game? How does each player modify his strategy from information about the other player's plan? In answering these questions we will begin by studying the zero-sum two-person game, in which each person may choose between two and only two distinct strategies.

52. Two-Person Zero-Sum Games. In introducing the methodology of game-theory analysis, it is well to begin with a very simple game. Consider, for example, that two players, A and B, are playing the game of matching pennies under rules whereby each player chooses the side of the penny he will show, and does not see, of course, the side shown by his opponent until the play is made. If the two faces match, Player A will win a unit sum from Player B; likewise, if

		Player B	
		Play #1 Heads	Play #2 Tails
Player A	Play #1 Heads	1	−1
	Play #2 Tails	−1	1

the two faces do not match, Player B will win a similar amount from Player A. This game may be represented by the following game matrix. This matrix is effective in summarizing the results of the various plays in the game.   Each row represents a play by Player A, each column a play by Player B, and the figures in the spaces formed by the intersection of rows and columns represent the payments for the plays.   We use the convention that positive numbers indicate a gain for Player A and consequently a loss for Player B, and that negative numbers indicate a gain for Player B and consequently a loss for Player A.

A *pure strategy* is a play that is available to a participant in a game. In the above example each player has two pure strategies at his disposal (i.e., he may show *either* one face of the penny or the other). Pure strategies are mutually exclusive in that, when a player selects one, he automatically forfeits the right of employing another on that particular play of the game.   A player's *grand strategy* is his method of selecting pure strategies for a series of plays in a given game.   A player arrives at his grand strategy by calculating the odds that indicate the most favorable mixture of his pure strategies.

The game of matching pennies seems to be a "fair" game, but let us now establish the mathematical criterion for such a supposition.   Our first step is to discover the odds that indicate the optimum grand strategy for each player.   To do this we first subtract the second row from the first row in the game matrix and write the results, after interchanging them, directly below the figures in the second row. Then we subtract the second column from the first column and write the results, again after interchanging them, directly to the right of the figures in the second column.   This gives us:

		Player B		
		1. Heads	2. Tails	
Player A	1. Heads	1	−1	−2
	2. Tails	−1	1	2
		−2	2	

Note that the −2 below the first column was obtained by disregarding the elements in the first column and by subtracting the elements in the second column as shown on the next page.

Player B

	1. Heads	2. Tails
Player A 1. Heads		-1
2. Tails		1

$(-1) - (1) = (-2)$

-2

The other differences were determined similarly by subtracting the appropriate elements and interchanging the results. The absolute values* of the four differences just obtained represent the odds or the ratio in which a player should mix his pure strategies. Thus Player A should select either of his two pure strategies at random, but all the while he must keep in mind that he should be using the two strategies in a 2 : 2 or 1 : 1† ratio over a period of time.

Likewise, Player B should select his strategies at random but in accordance with the 1 : 1 ratio. In other words, each player should decide by some device over which he has no control (such as flipping a coin) which pure strategy to select for a given play. By not using such a device, a player who seemingly makes an arbitrary decision for each play may be unconsciously following a system that can be interpreted by his opponent and used against him.

A "fair" game is defined as one whose value is zero. That is, if such a game is played over an extended length of time, neither player has a better chance than the other of reaping a profit. The value of a game can be found by selecting a row at random and multiplying each element through by the corresponding term of the odds affixed to the columns and then dividing the sum of the products thus obtained by the sum of the integers that represent the column odds. Similarly a column may be selected at random and the same value will be obtained by multiplying each element in it by the corresponding term from the set of row odds and dividing the sum of these products by the sum of the

*The absolute value of a number is defined in *Mathematics Dictionary*, edited by James and James, 2nd edition, (1959), Van Nostrand, as: ". . . its value without regard for sign; its numerical value. The number 2 is the absolute value for both +2 and −2."

†Any set of odds may be reduced to simpler figures by dividing every term in the set by a common factor.

integers composing the set of row odds.   Thus, in the example we have been using, if we select the first row, we obtain

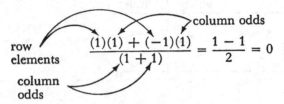

$$\frac{(1)(1) + (-1)(1)}{(1 + 1)} = \frac{1 - 1}{2} = 0$$

row elements

column odds

We will obtain the same result by selecting the remaining row or either column and conducting the appropriate operation.   Here the result is zero which proves that "matching pennies" is a fair game.   In any case the value of a game can be thought of as the average payment over a period of time, taken from the point of view of one of the players involved.   The applications of the operations just demonstrated will become more obvious when we now consider some games which possess values unequal to zero.

53. Saddle Points.   Let us now investigate games in which the players must employ preferred strategies.   An example of such a game is given in the following matrix, in which all the figures represent gains by A, and hence losses by B.

		Player B	
		1	2
Player A	1.	7	6
	2.	6	5

In this case, the numbers to the left of and above the matrix designate the pure strategies of Player A and Player B respectively.   As stated above, all the numbers within the matrix itself are positive and therefore represent gains for A and losses for B.   The main concern of Player A then is to find and adopt a grand strategy that will assure him of the maximum profit in spite of any grand strategy that Player B may adopt to thwart this purpose.   Likewise, the main concern of Player B is to find and adopt a grand strategy that will minimize his losses in spite of any grand strategy that A may adopt.

To find his optimum grand strategy, Player A first examines the rows of the game matrix (his pure strategies) and then writes to the right of the last figure in each row the lowest figure in that row.   These figures are called the *row* minima and represent the minimum gain available to Player A for each pure strategy at his disposal. Of these

	Player B		
	1	2	
Player A 1	7	6	⑥
2	6	5	5

he naturally prefers the pure strategy with the greatest minimum return, e.g. row 1 in this case.

Player B, on the other hand, examines the columns and lists below the last figure in each the largest figure in that column. These figures are called the *column maxima* and represent the maximum losses possible for Player B for each pure strategy at his disposal.

	Player B	
	1	2
Player A 1	7	6
2	6	5
	7	⑥

Of these he favors the pure strategy which entails the least maximum loss, e.g., column 2 in this case.   In this game we find that the maximum of the row minima (MAXMIN) is equal to the minimum of the column maxima (MINMAX).

	Player B		
	1	2	
Player A 1	7	*6	⑥
2	6	5	5
	7	⑥	

Whenever this occurs the game in question is said to have a *saddle point* and in such a case it is to the advantage of each player to select only that pure strategy which contains the saddle point. Thus Player A will always play his first pure strategy, i.e. the first row in the game matrix, and Player B will always play his second pure strategy, i.e. the second column in the game matrix, and Player A will make a consistent profit of 6 units per play. Here we see several things. First, a player's grand strategy is simply a pure strategy when there is a saddle point. Second, from our preceding definition of the value of a game, we can see that, in a game which contains a saddle point, the value of the saddle point is invariably the value of the game. Thus the value of this game to Player A is 6. In other words the average payoff is 6 units to Player A and a loss of 6 units to Player B. This is obviously not a fair game for the value of the game is not zero and the average payoff favors one player. Third, this game illustrates an important convention of game theory. The player who employs the row strategies is called the *maximizing player* and the player who employs the column strategies is called the *minimizing player*. This terminology does not indicate any basic difference between the outlooks of the two players. Although Player A seems to be playing to maximize his winnings and Player B seems to be playing to minimize his losses, there is essentially no disparity between the goals of the two players. The introduction of these two terms is, in fact, simply a consequence of adopting the convention of using positive numbers to represent the winnings of Player A and negative numbers to represent the winnings of Player B.

The observations that we have just made were all obtained by assuming that each player is intelligent and aware of the intelligence of his opposition. If either one of the above players selects the alternate pure strategy (the one not including the saddle point) he will only penalize himself in so doing. For example if Player A decides to select his second pure strategy while B maintains his second pure strategy, A will find his winnings reduced from 6 to 5 on that particular play. Likewise, if Player B decides to select his first pure strategy while Player A maintains his first pure strategy, B will find his losses have increased from 6 to 7 on that particular play. So it is safe to say that in any game which has a saddle point, the optimum grand strategies for the two players involved consist of playing only those pure strategies that include the saddle point and neglecting the others. This brings us to the first rule for finding the value of any game: *Always seek to establish whether the game in question has a saddle point before employing any of the other methods. If it does, this value is also the value of the game.* Incidentally, the game in the above example could be considered fair if Player A were to make a side payment of 6 units to Player B before each play. Thus any game, whose value can be determined, may be made fair by having the player whom the game favors make a side payment equal to the value of the game to that player who is otherwise at a disadvantage.

Example 1.   Find the saddle point in the following game.

Player B

		1	2
Player A	1	8	4
	2	1	2

First, we find the maximum of the row minima

Player B

		1	2	
Player A	1	8	4	④ MAXMIN = 4
	2	1	2	1

Second, we find the minimum of the column maxima.

Player B

		1	2	
Player A	1	8	4	MINMAX = 4
	2	1	2	
		8	④	

Third, we see that the MAXMIN is equal to the MINMAX. Thus there is a saddle point, the number 4, and the value of this game is then 4. Furthermore, Player A will select his first pure strategy each time and Player B will select his second.

54. Mixed Strategies. Now let us turn our attention towards some games that require the use of definite mixed strategies on the part of the players. Consider the game:

Player B

		1	2
Player A	1	4	−5
	2	−3	5

When we examine for a saddle point, we find

Player B

		1	2	
	1	4	−5	−5
Player A				
	2	−3	5	⊝3
		④	5	

MAXMIN = −3
MINMAX = 4
MAXMIN ≠ MINMAX
∴ no saddle point.

The next step then is to find the odds that describe the optimum mixture of the pure strategies for each player.   This we do by following the method outlined in our discussion of the game of matching pennies.

The odds relating B's pure strategies are found by subtracting the elements in the second row from the corresponding ones in the first and interchanging the results,

| Player B | | | Player B | | | Player B | |
1	2		1	2		1	2
.	−5		4	.		4	−5
.	5		−3	.		−3	5
−10	.		.	7		−10	7

Thus when we take the absolute values of these results we find that player B should play his first pure strategy 10 times for every 7 times he plays his second.   His optimum grand strategy, therefore, is a 10 : 7 random mixture of his two pure strategies.

The odds relating A's pure strategies are found by subtracting the elements in the second column from those in the first and interchanging the results.

$$\text{Player A} \begin{array}{c} 1 \\ 2 \end{array} \begin{array}{|c|c|} \hline . & . \\ \hline -3 & 5 \\ \hline \end{array} \begin{array}{c} -8 \\ . \end{array} \quad \text{Player A} \begin{array}{c} 1 \\ 2 \end{array} \begin{array}{|c|c|} \hline 4 & -5 \\ \hline . & . \\ \hline \end{array} \begin{array}{c} . \\ 9 \end{array} \quad \text{Player A} \begin{array}{c} 1 \\ 2 \end{array} \begin{array}{|c|c|} \hline 4 & -5 \\ \hline -3 & 5 \\ \hline \end{array} \begin{array}{c} -8 \\ 9 \end{array}$$

Taking the absolute values of these results, we find that Player A should play his first pure strategy 8 times for every 9 times he plays his second.   Consequently, his optimum grand strategy is a 8 : 9 random mixture of his two pure strategies.

To find the value of this game we select any row or column, and proceed according to the following equations to crossmultiply, add, and divide:

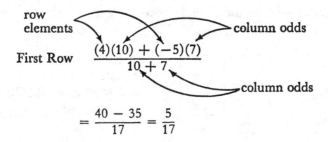

$$= \frac{40 - 35}{17} = \frac{5}{17}$$

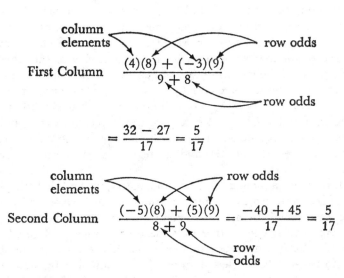

row
elements   column odds

Second Row $\dfrac{(-3)(10) + (5)(7)}{10 + 7} = \dfrac{-30 + 35}{17} = \dfrac{5}{17}$

column odds

column
elements   row odds

First Column $\dfrac{(4)(8) + (-3)(9)}{9 + 8}$

row odds

$= \dfrac{32 - 27}{17} = \dfrac{5}{17}$

column   row odds
elements

Second Column $\dfrac{(-5)(8) + (5)(9)}{8 + 9} = \dfrac{-40 + 45}{17} = \dfrac{5}{17}$

row
odds

Here we have worked out the values using each column and each row. In every instance we see that each element in a row or column was multiplied by the absolute value of corresponding integer from the set of column or row odds respectively, and then the sum of these products was divided by the sum of the integers representing the same odds. It is seen that the value of this game is $\frac{5}{17}$ which represents the average amount won per play by Player A when both players are using their optimum mixed strategies. The only precaution that Player A must observe in order to ensure a continual profit of $\frac{5}{17}$ is to keep secret the pure strategy that he intends to use on each future play. He may permit Player B to know all of his past decisions as well as to know his grand strategy. If it is a good grand strategy, Player B can in no way prevent Player A from obtaining the average payment due him. Likewise, Player B must maintain the same precautions or his losses will be greater.

The only foolproof method of deciding which pure strategy to employ on a given play is to let the decision rest entirely on some chance event. Thus a chance mechanism is an indispensable part of a good grand strategy. In the problem that we have just worked out, Player A might select his pure strategies by putting into a bowl 8 orange and 9 black marbles, all of the same size, shape, weight, and texture, and then selecting one of them with his eyes closed. If he should draw an orange marble, then he would play his first strategy; a black marble, his second strategy. Before drawing for each successive play, Player A must return the marble he drew for his present play so that the bowl always contains the same number of marbles in the same ratio. The best method, however, is to employ a table of random numbers such as the one shown on page 00. For example, using this table to mix two strategies, a and b, in a 3 : 1 ratio, we would pick a starting point at random and then run down each column, playing strategy a for all numbers between 1 and 150 and strategy b for all numbers between 150 and 200.

Returning for a moment to the game matrix, we find that the play of a game is not altered by adding a constant to each payoff or by multiplying each payoff by a positive constant. However the value of the game is affected when either of these operations is performed. For the game that we analyzed previously in this section, we found the value to be $\frac{5}{17}$. Note the values obtained below first by adding 2 to each term (a) and then by multiplying each term by 3, (b).

(a)

6	−3
−1	7

$$\text{Value} = \frac{(6)(10) + (-3)(7)}{17} = \frac{39}{17} = 2\frac{5}{17} = 2 + \frac{5}{17}$$

(b)

12	−15
−9	16

$$\text{Value} = \frac{(12)(10) + (-15)(7)}{17} = \frac{120 - 105}{17} = \frac{15}{17} = 3\left(\frac{5}{17}\right)$$

Hence, when a constant is added to each element in the game matrix, the value of the game is also increased by that number. Similarly, when each element in the game matrix is multiplied by a positive constant, the value of the game is multiplied by the constant. In most cases the play as well as the value of a game would be influenced by multiplying each element in the game matrix by a negative constant.

Now let us examine two problems that will demonstrate how the techniques of game theory may be applied and how a stated problem may be translated into the notation of game theory.

Prob. 1. Recall the game of matching pennies. If strategy 1 for either player means that he shows heads and if strategy 2 means that he shows tails, which of the following game matrices illustrates the more advantageous set of payoffs for Player A?

I.

		Player B	
		1	2
Player A	1	3	0
	2	−4	−1

II.

		Player B	
		1	2
Player A	1	−2	3
	2	1	−2

To answer this question we must find the value of each of the above games. The game having the higher numerical value will be more advantageous to Player A. Let us examine matrix I first.

Our first step is always to check for a saddle point, so we begin by finding and comparing the maxmin and the minmax.

		Player B	
		1	2
1		3	0
2		−4	−1
		3	⓪

⓪ MAXMIN = MINMAX

−4 ∴ There is a saddle point at zero.

Thus we have already found the value of matrix I to be zero. Checking for a saddle point in matrix II we find

		Player B	
		1	2
Player A	1	−2	3
	2	1	−2
		①	3

(− 2) MAXMIN ≠ MINMAX

(− 2) ∴ no saddle point

Consequently we must compute the odds for the two sets of strategies before we can determine the value of the game.

		Player B		
		1	2	
Player A	1	−2	3	3
	2	1	−2	5 *
		5	3	

*The minus signs preceding some of the terms in a set of odds are not necessary once the entire set has been determined.

Selecting the first row, we find the value of matrix II.

$$\frac{(-2)(5) + (3)(3)}{(5 + 3)} = \frac{-10 + 9}{8} = -\frac{1}{8}$$

Thus player A would experience an average loss of $\frac{1}{8}$ unit per play if both players used their optimum grand strategies (a mixture of 5 : 3 for A and 3 : 5 for B). This is worse than just breaking even; thus we have found that matrix I is more advantageous to player A.

Prob. 2. A certain business buys and sells two commodities, commodity A and commodity B. In general, there are only two marketing conditions that it must consider: condition X and condition Y. It has no control over these conditions which may alternate at random, but it does know that under condition X, it can sell 60 items of commodity A per week, and under condition Y it can sell 10 items of commodity A and 100 items of commodity B per week. Commodity A costs the business concern $2 per item but it can be resold for $5. Commodity B costs $1 but can be resold for $2. The business concern is willing to invest $120 in goods at the beginning of each week, but at the end of the week, all the merchandise that has not been sold is considered a total loss, for both commodities are perishable. In what proportion should the business concern buy for condition X and for condition Y and what is its optimum average income per week?

We construct the following matrix

		Sells for	
		1.	2.
		Condition X	Condition Y
Purchase for	1. Condition X	180	−70
	2. Condition Y	−70	130

These figures represent the business concern's potential losses and gains under each given condition. For example, if it buys for condition X, it invests all of the $120 in commodity A. If, however, condition Y should then prevail, the concern is able to dispose of only 10 of the 60 items and thus experiences a loss of $70 (upper right-hand corner of matrix).

First we check for a saddle point and find that there is none. Then we must find the odds that will provide the concern with the optimum grand strategy.   To simplify the calculations, we divide every element in the game matrix by 10 and then calculate the odds.   This gives us:

		Sells for		
		1. X	2. Y	
Purchases for	1. X	18	−7	20
	2. Y	−7	13	25
		20	25	

Thus the concern should buy for condition X and condition Y according to the odds 4 : 5.   The value of the game is

$$(10^*)\left[\frac{(18)(4) + (-7)(5)}{4 + 5}\right] = (10)\frac{37}{9} = \frac{370}{9} = \$41.11$$

As an alternative to playing the odds, the business concern could decide to invest $\frac{4}{9}$ of the \$120 in merchandise for condition X and $\frac{5}{9}$ in merchandise for condition Y.   It would then purchase \$64.50 worth of commodity A and \$55.50 worth of commodity B.   This investment would ensure a steady profit of \$42.11 in spite of the fluctuation in the prevailing condition.   It is not always possible to form a synthesis of the strategies as we did here, for there are many cases that call for the selection of one and only one of a series of mutually exclusive pure strategies per play.

55. 2 × N Games.   Games in which one player has two strategies and the other has more than two are described as '2 × n' (two by n) games.   Here the n may be replaced by any integer greater than two. As we shall soon see, only a few new techniques must be introduced to describe these games as an extension of the 2 × 2 games.

In every case a 2 × n game will have either a saddle point solution or an optimum mixed strategy based on only two pure strategies. *Thus, any 2 × n game may be reduced to an equivalent 2 × 2 game.*

As emphasized in the previous chapter, the first step in analyzing any game is to look for a saddle point.   If there is one, the game is solved and some unnecessary calculation has been avoided.   The technique for finding a saddle point for 2 × n games is the same as that used for 2 × 2 games.   We simply find and compare the maxmin and the minmax.   Consider the following game.

*Remember that previously we divided the game matrix through by 10.

Player B

		1	2		
	1	6	5	5	
	2	5	5	5	
Player A	3	6	4	4	row minima
	4	7	6	⑥	
	5	2	4	2	

column maxima   7   ⑥

MAXMIN = MINMAX = 6

Thus, this game does have a saddle point and we know immediately that Player A's optimum grand strategy is his fourth pure strategy and that Player B's optimum grand 'strategy is his second pure strategy. In other words, A plays his pure strategies according to the odds 0 : 0 : 0 : 1 : 0 and B plays his pure strategies according to the odds 0 : 1. The value of this game is obviously 6.

If, however, the given 2 × *n* game does not contain a saddle point, the next step is to attempt to reduce it to a 2 × 2 matrix. When we study the choices of the player who has many strategies at his disposal, it may be obvious that some of them are inferior to others. In this case the superior strategies are said to *dominate* the inferior ones and the latter may be eliminated from the game. For our present purposes, there are two types of dominance. When the values of all the elements in one pure strategy are greater than those of the corresponding elements of another, the situation is defined as *strict dominance*. However, when some of the values are equal, the situation is defined as *non-strict dominance*. It is always safe to eliminate a strategy on the basis of strict dominance. However, elimination of a strategy on the basis of non-strict dominance may cause a subtle alteration in a game that has more than one solution. Such games as these are of a higher order than we have yet discussed. The concept of dominance is an important one in game theory and will be elaborated on further in the subsequent sections of this chapter. Let us now consider the game at the top of the next page.

If we check for a saddle point, we find that there is none. Our next step is to try to single out A's dominant pure strategies. Each element in A's fourth strategy is seen to be greater than the correspond-

Player B

	1	2
1	0	3
2	2	1
Player A 3	1	4
4	3	2
5	2	1

ing elements in his second and fifth strategies. Thus, we may eliminate both of these from the game matrix. This gives us:

Player B

	1	2
1	0	3
Player A 3	1	4
4	3	2

It is now evident that A's third strategy is dominant over his first strategy, so we eliminate the latter and obtain

Player B

	1	2
Player A 3	1	4
4	3	2

This game may now be solved by the methods we have described for operations with 2×2 games. Thus, we find that A should play his third and fourth strategies according to the odds 1 : 3 and that B should play his first and second strategies according to the odds 1 : 1. The value of the 2×2 game, and consequently of the $2 \times n$ game, is 2.5. Player A may consider his set of odds for the original game to be 0 : 0 : 1 : 3 : 0. The usefulness of dominance is quite evident in this example. It may also be applied to simple 2×2 games, but its occurrence in them indicates that the game must have a saddle point.

The next situation that must be discussed is the $2 \times n$ game that does not contain a saddle point and cannot be reduced by dominance to a 2×2 game. After we have examined a game and found it without a saddle point, it may still be possible to reduce the game somewhat by dominance. If A is the multistrategy player, we attempt to reduce the game by eliminating the *dominated* strategies. However, if B is the multistrategy player, we attempt to reduce the game by eliminating the *dominant* strategies. This follows, for A is striving to maximize his profits and B is striving to minimize his losses. After the game has been reduced as much as possible by dominance, the next step is to take one of the remaining 2 × 2 games, solve it, and test the solution in the original game. That is, first solve the 2 × 2 game and then, using the odds proposed for the two-strategy player, see how this player does against each of the strategies of the multistrategy player. If the two-strategy player does *as well or better* against the strategies not in the 2 × 2 subgame then the solution has been discovered. As we noted earlier, a player with a good grand strategy can win consistently against either of his opponents' pure strategies. This remark may now be extended. In the 2 × *n* game, one can win the same amount against either of the strategies that form the opponent's best mix. However, one wins a greater amount against the opponent's remaining strategies. Occasionally the winnings are the same against the remaining strategies but this is the exception and not the rule. Consider the following game.

		Player B						Row
		1	2	3	4	5	6	Min.
Player A	1	−7	−2	0	3	3	2	−7
	2	6	−3	5	2	−6	6	−6
Col. MAX.		6	−2	5	3	3	6	

MAXMIN = −6   MINMAX = −2

MAXMIN ≠ MINMAX

∴ no saddle point.

Upon further investigation we discover that B's third, fourth, and sixth strategies dominate his second strategy so these may be eliminated, to give us:

		Player B		
		1	2	5
Player A	1	−7	−2	3
	2	6	−3	−6

There is no dominance among B's remaining three strategies. Consequently we must search for a 2 × 2 subgame within the 2 × 3 game which will satisfy the latter (thus automatically satisfying the original 2 × 6 game). We begin with

Player B

		1	2
Player A	1	−7	−2
	2	6	−3

and find that Player A should play according to the odds 9 : 5 and Player B should play according to the odds 1 : 13. Calculating the value of this subgame, we find

$$\frac{(9)(-7) + (5)(6)}{9 + 5} = \frac{-33}{14}$$

To see if this solution will satisfy the 2 × 3 game we must test A's 9 : 5 strategy on B's remaining pure strategy (B-5). If the result is greater than or equal to −33/14, the 2 × 3 game is satisfied.

$$\frac{(9)(3) + (5)(-6)}{14} = \frac{-3}{14}$$

This result satisfies our criterion. Consequently the 9 : 5 mixture is a good strategy for Player A and the 1 : 13 : 0 : 0 : 0 : 0 mixture is a good strategy for Player B.

Let us compare these results with those we obtain by using another 2 × 2 subgame of the given 2 × 3 game.

Player B

		2	5
Player A	1	−2	3
	2	−3	−6

We find that thus subgame has a saddle point at −2. Thus, according to this matrix, A should always select his first strategy and B should always select his second. But now we must test A's 1 : 0 mix on B's first strategy and we find that we obtain the value −7 which is considerably less than −2. Therefore, this subgame does not satisfy our criterion for a good strategy and its solution does not satisfy the 2 × 3 or 2 × 6 game matrix.

It is imperative to reduce a 2 × n game as much as possible before starting the search for the 2 × 2 solution, for there are $\dfrac{n(n - 1)}{2}$ games of the 2 × 2 variety contained within every 2 × n game. If we had not reduced the above 2 × 6 game, we would have been faced with the analysis of 15 possible 2 × 2 subgames.

Another very practical method of solving 2 × n games consists of graphing the various strategies that are at the disposal of the multi-strategy player. Recall the game matrix

		Player B					
		1	2	3	4	5	6
Player A	1	−7	−2	0	3	3	2
	2	6	−3	5	2	−6	6

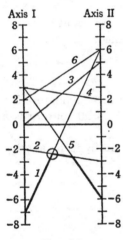

Fig. 4-1.

This game may be graphed as shown in Figure 4-1 by plotting the first element in each of B's pure strategies on one vertical axis and by plotting the second element in each of B's pure strategies on a second vertical axis and by then connecting each of the six sets of corresponding points with straight lines. Next the line segments that bound the bottom portion of the graph (since B is the minimizing player) are marked with a heavier line. The highest point on this heavy line marks the intersection of the combination of pure strategies that will satisfy the original game. Here we see that the point lies on the intersection of the first and second pure strategies, which we have already proven correct by computation. This method of graphing provides a useful shortcut for determining which 2 × 2 subgame is the solution of a given 2 × n game. Once the desired subgame has been found, it is an easy matter to compute the odds and value attached to the original game.

If the game in question provides Player A (the maximizing player) with the greater number of strategies, a similar graph may be drawn, but the line segments bounding the top rather than the bottom of the figure are now relevant and the lowest point on this broken line marks the intersection of the preferred mixture of strategies. Note the graphing of the following game.

Player B

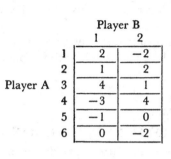

		1	2
	1	2	−2
	2	1	2
Player A	3	4	1
	4	−3	4
	5	−1	0
	6	0	−2

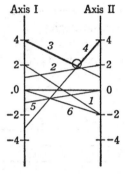

Fig. (4-2)

From this graph we can assume that A's best mixture of pure strategies consists of his third and fourth alternatives. This seems a little strange for his fourth strategy has the greatest loss potential of any of his pure strategies. Why wouldn't a mixture of his second and third pure strategies be more profitable, for both of these preclude the possibility of loss? The answer is that, in this case, it is more profitable to gamble a little and play the odds. First let us solve the 2 × 2 subgame which contains A's third and fourth pure strategies. We find that A should play these two strategies in the ratio 7 : 3 and that B should play his first and second pure strategies in the ratio 3 : 7. The value of this game for A is 1.9 and it conforms with our criterion for a solution; i.e., when B's odds are tried on A's remaining strategies, the result is invariably less than the value of this 2 × 2 subgame. For example, when B's 3 : 7 odds are tried on A's second strategy, we obtain 1.7.

Likewise, if we solve the 2 × 2 subgame which contains A's second and third pure strategies, we find that A could best play these according to the ratio 3 : 1 while B plays his first and second strategies according to the ratio 1 : 3. However, the value of this game for A is merely 1.75. Thus, a little daring and a sound grand strategy will provide more of a profit in this case than will an untested selection of the two strategies that contain no element of risk.

56. 3 × 3 Games. The games that will be discussed in this section are similar to those already described in that we are still dealing with two-person zero-sum games, i.e. games that can be played by only two competitors with the sum of one player's gains equal to the sum of the other player's losses.

The difference here is that each player has at his disposal three instead of two strategies.

Note as before that the first step in evaluating any game is to determine the maxmin and minmax and look for a saddle point. Consider the following game.

		Player B		Row Min.	
	1	2	3		
1	2	1	0	⓪	MAXMIN = MINMAX = 0
Player A 2	−4	−3	−2	−4	∴ There is a saddle point.
3	3	4	−1	−1	
Col. Max.	3	4	⓪		

Thus the game is immediately solved.   The value is 0 and A's optimum grand strategy is $1 : 0 : 0$ while B's is $0 : 0 : 1$.   If either player deviates from the saddle point strategy, he and he alone will suffer for it.

Just as the saddle point technique is seen to apply in the study of 3×3 games, so may the idea of dominance be extended in a similar manner.   Consider the game

		Player B		
		1	2	3
Player A	1	−3	1	−1
	2	7	1	−2
	3	1	7	2

At once we see that A's third pure strategy dominates his first.   The elimination of the latter reduces this matrix to a 2×3 game

		Player B		
		1	2	3
Player A	2	7	1	−2
	3	1	7	2

next we notice that B's second strategy dominates his third.   When the former is eliminated we obtain

		Player B	
		1	3
Player A	2	7	−2
	3	1	2

This 2×2 game, when solved, will provide the optimum grand strategies and the value for the original game. If we are confronted with 3×3 game that does not have a saddle point, the next step is to try to reduce it by dominance to a $2 \times n$ game, which can always be further reduced to a 2×2 game.

In the above game A should follow the grand strategy 0 : 1 : 9 while B should follow a 2 : 0 : 3 mixture. The value of the 2×2 game, and consequently of the 3×3 game, is 1.6.

Of course, there are 3×3 games which have neither a saddle point nor dominant strategies and there is a method which may be employed successfully on most games falling into this category. The easiest way to introduce this approach is to show its application to a specific example. Consider the following:

		Player B	
	4	−1	0
Player A	0	1	3
	−2	4	1

By inspection we find that this game has no saddle point and no dominant strategies. Thus we must compute the odds governing the optimum grand strategies of each player. First subtract each element in the second row of the game matrix from the corresponding element in the first row and write the results directly beneath the elements in the third row. Then subtract each element in the third row from the corresponding element in the second row and write these results directly beneath the last row of figures. Now we have

			Player B	
		1	2	3
	1	4	−1	0
Player A	2	0	1	3
	3	−2	4	1
		4	−2	−3
		2	−3	2

To find the set of 3 integers that represents B's optimum grand strategy we proceed as follows. The integer associated with B's first strategy is found by eliminating the two numbers (4, 2) appearing below the first column and computing the determinant of the square matrix formed by the remaining numbers. The absolute value of this determinant is the required integer. The operation is performed below:

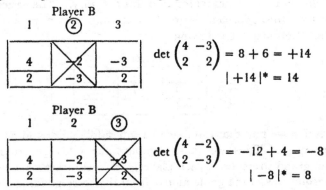

Player B

	①	2	3
		−2	−3
	2	−3	2

$$\det \begin{pmatrix} -2 & -3 \\ -3 & 2 \end{pmatrix} = -4 - 9 = -13$$

$$|-13|^* = 13$$

Likewise to find the integers associated with B's second and third strategies we perform the following operations.

Player B

	1	②	3
	4	2	−3
	2	−3	2

$$\det \begin{pmatrix} 4 & -3 \\ 2 & 2 \end{pmatrix} = 8 + 6 = +14$$

$$|+14|^* = 14$$

Player B

	1	2	③
	4	−2	3
	2	−3	2

$$\det \begin{pmatrix} 4 & -2 \\ 2 & -3 \end{pmatrix} = -12 + 4 = -8$$

$$|-8|^* = 8$$

At this point we have determined B's optimum grand strategy to be a 13 : 14 : 8 mixture of his pure strategies. We may find A's optimum grand strategy by performing a similar operation.

First subtract each element in the second column from the corresponding one in the first and write the results directly to the right of the figures in the third column. Next subtract each element in the third column from the corresponding one in the second and write these results to the right of those just obtained.

Player B

		1	2	3		
	1	4	−1	0	5	−1
Player A	2	0	1	3	−1	−2
	3	−2	4	1	−6	3

*The symbol used here may be translated as "the absolute value of," e.g. $|-13| = 13$ is stated as "the absolute value of -13 is 13". The concept of absolute value was defined in a footnote near the beginning of this chapter.

To determine the integer associated with A's first strategy, strike out the two differences obtained from the elements in the first row and find the determinant of the matrix formed by the four remaining differences. The absolute value of this determinant is the required integer. Thus we have:

Player A   ①   2   3

~~5~~	~~−1~~
−1	−2
−6	3

$$\det \begin{pmatrix} -1 & -2 \\ -6 & 3 \end{pmatrix} = -3 - 12 = -15$$

$$|-15| = 15$$

Similarly, to obtain the remaining integers, we perform the following operations

Player A   1   ②   3

5	−1
~~−1~~	~~−2~~
−6	3

$$\det \begin{pmatrix} 5 & -1 \\ -6 & 3 \end{pmatrix} = +15 - 6 = +9$$

$$|+9| = 9$$

Player A   1   2   ③

5	−1
−1	−2
~~−6~~	~~3~~

$$\det \begin{pmatrix} 5 & -1 \\ -1 & -2 \end{pmatrix} = -10 - 1 = -11$$

$$|-11| = 11$$

Consequently A's optimum grand strategy is a 15 : 9 : 11 mixture of his three pure strategies. In performing the above operations, it is important to note that all + and − signs were maintained throughout the processes of subtraction and finding the determinants. No signs may be dropped in any of this computation until all the determinants have been calculated. Then the absolute values of the determinants are taken and these values then compose the set of odds describing a grand strategy.

To find the value of a 3 × 3 game, we simply extend the method used in solving those of the 2 × 2 variety. For example, we may select the first row of the above game matrix and multiply each element in it by the corresponding integer from B's grand strategy. The sum of these products divided by the sum of the integers representing B's grand strategy is the value of the game. This operation is conducted below.

A's first row   4   −1   0

B's grand strategy   13   14   8

$$\frac{(4)(13) + (-1)(14) + (0)(8)}{(13 + 14 + 8)} = \frac{52 - 14}{35} = \frac{38}{35}$$

Some 3×3 games defy solution by any of the three methods described so far. For example, a 3×3 game may appear perfectly regular and have no saddle point or dominant strategies. But when we attempt to calculate the grand strategies by the subtraction-determinant method, we find in certain cases that the resulting strategy has the form $0 : 0 : 0$, which is an indeterminate solution. To find a correct solution for this type of game, we have no choice but to resort to the trial and error method. This entails the elimination of one strategy at random and the solution of the remaining 2×3 game (which of course can be reduced to a 2×2 game). As in the method we used for solving $2 \times n$ games, the grand strategy derived from a given subgame must be tested by applying it to the eliminated strategy. If the solution of the subgame is in accordance with the criterion that we stated in the section on $2 \times n$ games, it is also a solution of the original game. Consider the following game.

		Player B		
		1	2	3
	1	4	−6	−1
Player A	2	2	−4	−1
	3	0	2	1

This game has no saddle point and no dominant strategies. When we attempt to calculate the grand strategies by the determinant method, we find that B should play according to a $1 : 1 : 2$ ratio and that A should play according to a $0 : 0 : 0$ ratio. This is not a determinate solution so we must now employ the trial and error method.

First let us drop A's third strategy and solve the remaining 2×3 game. We have

		Player B		
		1	2	3
Player A	1	4	−6	−1
	2	2	−4	−1

This game has a saddle point at $−4$. Consequently this suggests that the value of the original game is $−4$ and that A's optimum grand strategy is $0 : 1 : 0$ while B's is also $0 : 1 : 0$. We now test B's strategy against A's third strategy and obtain the result $+2$. This indicates that there is a flaw in the grand strategies that we acquired from the subgame, for if B plays the strategy assigned to him and A changes his strategy to a 0 : 0 : 1 mix, A will profit.   A player who deviates from his optimum grand strategy can not profit; therefore the strategies that we calculated here were spurious.   Let us now try the 2 × 3 subgame formed by the elimination of A's second strategy.   We have:

Player B

		1	2	3
Player A	1	4	−6	−1
	3	0	2	1

This game has no saddle point and no dominant strategies.   If we attempt to find B's best mixture of 2 strategies by the graph method, we find that all three lines intersect at the same point.   This phenomenon indicates that there is more than one acceptable solution to the game.   Consequently, we must now select any one of the 2 × 2 games included within this 2 × 3 game and hope that its solution will satisfy the original game.   If we eliminate B's third strategy, we have:

Player B

		1	2
Player A	1	4	−6
	3	0	2

This game has no saddle point and no dominant strategies.   Upon solving it, we find that A should play his two pure strategies in the ratio 1 : 5 and B should play his two in the ratio 2 : 1.   The value of the game is found to $\frac{2}{3}$.   When we test A's mixed strategy against B's third strategy, we also obtain $\frac{2}{3}$.   So far, this result fits the criterion for a solution.   Now we must test these strategies in the original 3 × 3 game.   The 3 × 3 game and its odds appear as:

Player B

Player A	4	−6	−1	1
	2	−4	−1	0
	0	2	1	5
	2	1	0	

If A should desert his 1 : 0 : 5 strategy and play his second pure strategy while B maintains his 2 : 1 : 0 mixture, A would reduce the value of the game from $\frac{2}{3}$ to zero.   This is not to his advantage so he stands by his 1 : 0 : 5 strategy.   If B should desert his 2 : 1 : 0 strategy and play his third pure strategy while A maintains his 1 : 0 : 5 mix, the value of the game remains unchanged. Thus B can do nothing to prevent A from gaining a minimum average profit of $\frac{2}{3}$. However, B does have an alternative strategy at his disposal. He may play his pure strategies according to the ratio 1 : 0 : 2 instead of 2 : 1 : 0. This does not greatly benefit him for the value of the game remains $\frac{2}{3}$.

In most instances, a 3 × 3 game may be solved without resorting to this trial and error method. The example that we have just worked out is a special case which demonstrates the occasional limitations of our other methods.

57. Conclusion. In the foregoing sections of this chapter we have developed a number of methods for attacking simple games. There are no rigid restrictions on the dimensions that a game may have, but, as the matrices become larger, then solutions become proportionately more difficult. The techniques that we have employed in the solution of 2 × 2 and 3 × 3 games can be extended to apply to games of a higher order. However, in many cases, the mathematician must resort to computational methods that are beyond the scope of this book or must rely upon some appropriate mechanical device.

In solving any two-person zero-sum game, regardless of the number of strategies available to the players, our approach should follow this pattern: (1) Look for a saddle point; (2) Eliminate the meaningless strategies by examining for dominance; (3) Attempt the direct computation of the odds; (4) Trial and error. A few additional remarks about these methods may be helpful. Nothing else needs to be said about saddle points but the introduction of one other type of dominance may be of interest. A strategy may be dominated by a *combination* of other strategies. Consider the game:

			Player B	
		1	2	3
	1	7	1	4
Player A	2	1	4	2
	3	0	8	3

There is no saddle point and there is no dominance according to our original criterion. However, if each of A's strategies is considered as a unit, then strategy A-1 plus strategy A-3 dominate a double portion of strategy A₂. In other words, if the corresponding elements of A's first and third strategies are added, these totals dominate twice the value of the elements in A's second strategy.

A's First Strategy	7	1	4
+ A's Third Strategy	0	8	3
totals	7	9	7
2(A's Second Strategy)	2	8	4

Each element of the 7, 9, 7 combination dominates the corresponding element of the 2, 8, 4 combination. Consequently A's second strategy may be eliminated and the game matrix may be reduce to:

		Player B		
		1	2	3
Player A	1	7	1	4
	3	0	8	3

This type of dominance is frequently useful in the simplification of complex games.

The subtraction determinant method may also be extended to higher order square games. However, the determinants can be quite involved and it is frequently necessary to simplify them by employing our knowledge of determinant properties.

Until now we have been concerned solely with two-person games. In closing this chapter a few remarks should be made about the complexities of games that have three distinct parties in competition.

A three person game introduces many complications which do not arise in the two-person game. These complications stem from the possibility of coalitions, that is, a situation in which two of the players act in union to play against the third player. From game theory it follows that for players A, B, and C, there are four possible situations: (1) A three-person game between A, B and C; (2) A two-person game between A, B on one hand and C on the other; (3) A two-person game between the A-C combination and the player B; and (4) A two-person game between the B-C combination and the player A. Since the returns to the players will in all probability be different for all four of these games, the analysis of three-person games is far more complex than that of two-person games and the complexity increases, of course, with increasing numbers of players until, as indicated at the beginning of the chapter, that number reaches a point where the effect of each player is negligible.

The theory of games assumes that coalitions will be formed whenever they yield greater advantages to two of the players than they would obtain if each acted independently. Matrices for the three-person games can also be set up. Usually one of several simplifying assumptions is made, one of them being that gains for each individual player are possible only if he forms a coalition with another player. In such cases the entire objective of the game for each player is to form a coalition with another. The solutions of such games consist, therefore, in determining which of the three possible coalitions yields the greatest return to its two members. A further complication of such games, however, is the possibility of one of the members of the coalition dealing with the third player. For example, if A and B form a coalition whereby each profits by one unit at the expense of C who loses two units, then A might deal with C on the basis that C pays A $1\frac{1}{2}$ units in all, on condition of A's withdrawal from the coalition with B — in this manner C would reduce his potential loss to $1\frac{1}{2}$ units, while A would extend his gain from the one unit to be obtained from his coalition with B, to the $1\frac{1}{2}$ units he would obtain directly from C.

To deal with this situation the concept of dominance may be extended from two-person games to three-person games. Thus, one of the possible distributions of payments in a three-person game is said to dominate another if it is more advantageous to both the members of the coalition (in the above example to both A and B).

It is apparent at once, however, that the extension of the concept of dominance to three-person games does not exclude multiple solutions, that is multiple distributions of payments which are of equal value. Assuming, for example, that the maximum payment which can be obtained by two players acting in coalition from the third player is the same whether the third player be A, B or C, then the theory of games obviously will not return a single answer as to which players should form the coalition. In cases of this sort, especially as they occur outside of the field of elementary games, in real games, or more particularly, in economics or other fields of human activity, there are often external conditions which limit the choices by the players to a smaller number than that mathematically possible. Thus, in the field of business there is specific legal prohibition of certain forms of cartel action.   Moreover, the moral or social effect of this legal prohibition is to discourage even smaller businesses, to which it does not apply directly, from combining for the complete exploitation of a third enterprise.   Their actions of this nature result far more frequently from the independent discovery by each in his own enterprise and from the results of his own operations, of a course of action which promises to yield the maximum profit, or the maxmin profit.

Problems for Solution

1. Find the optimum grand strategies and the values for the following 2 × 2 games:

(a)

		B 1	2
A	1	−6	11
	2	1	−6

(b)

		B 1	2
A	1	3	−1
	2	2	−4

(c)

		B 1	2
A	1	−12	12
	2	7	−7

2. Solve the following 2 × n games:

(a)

		B 1	2	3
A	1	5	0	2
	2	3	−2	−7

(b)

		1	2	B 3	4	5	6
A	1	7	0	19	9	10	13
	2	19	13	12	6	7	13

(c)

		1	B 2	3
A	1	1	3	−3
	2	0	−3	4

3. Solve the following 3 × 3 games:

(a)

		B		
		1	2	3
	1	−5	11	0
A	2	−3	−2	0
	3	2	12	3

(b)

		B		
		1	2	3
	1	−2	7	−5
A	2	9	−5	−6
	3	−3	2	0

(c)

		B		
		1	2	3
	1	7	4	−1
A	2	−9	7	2
	3	0	−5	1

(d)

		B		
		1	2	3
	1	9	−6	−5
A	2	−7	5	4
	3	−4	0	5

Chapter 5

INEQUALITIES, LINEAR PROGRAM-MING, AND THE TRANSPORTATION PROBLEM

58. Introduction. At first thought mathematics seems primarily concerned with the study and derivation of equalities. If we consider any of a number of branches of mathematics — e.g. arithmetic, algebra, geometry, trigonometry, or calculus, we find that this generalization can be justified to a substantial extent.

However, the study of inequalities, though less familiar to the non-mathematician, is also a highly significant and flourishing branch of mathematics. One of the most important aspects of the study of inequalities is that it provides a basis for the analysis of a class of problems that deal with maximization or minimization. Thus besides having a direct application to many geometrical problems, the study of inequalities has gained increasing recognition for its usefulness in solving practical problems for business and industry, as well as science and engineering.

The problem of minimizing or maximizing a linear function (e.g. costs, profits, etc.) subject to restrictions in the form of linear inequalities is treated by a comparatively new method, that of linear programming. Regarded as one of the most important applications of the theory of inequalities, linear programming can be used to solve a host of problems that at one time were considered beyond the methods of mathematics, or at any rate, of applied mathematics. Later in this chapter we elaborate on some of the aspects of linear programming and demonstrate how its techniques may be used to solve a variety of problems.

However, before giving applications of inequalities, it is necessary to devote some space to an examination of the fundamentals. Consequently, the first part of this chapter introduces the notation and the concepts that are axiomatic for a study of this branch of mathematics. Much of this material may seem quite self-evident or may already be familiar but the brief presentation of these fundamentals here is of value in understanding other parts of this chapter, and gives a better appreciation of the concepts underlying much of modern mathematics.

59. Fundamental Symbols and Axioms. The two most fundamental symbols in the notation of inequalities are " > ," which is translated as "greater than" or "is greater than," and " < ," which is translated as "less than" or "is less than." These symbols are used frequently in the study of geometry to compare line segments, angles, etc. But more generally these symbols are used when comparing any two real numbers. We may consider a real number to be any rational or irrational number. Geometrically a real number can be defined as any number lying on one of the axes defining the xy plane. Figure 5-1 shows a portion of the X-axis which may be considered to extend indefinitely to the left and to the right.

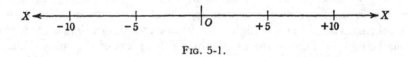

FIG. 5-1.

Any number on this axis may be considered greater than any number that is to its left. Thus, in the notation of inequalities we can say

$$0 > -5 \qquad \frac{1}{2} > \frac{1}{3} \qquad -\frac{1}{3} > -\frac{1}{2}$$

Expressed in slightly different terms, real number a is greater than any other real number b provided that the expression $a - b$ is a positive number. Thus, $a > b$ if and only if $(a - b) > 0$.

There are two axioms which are of fundamental importance in developing the theory of inequalities.

Axiom I — If a is a real number, it must satisfy one and only one of the following statements: a is equal to zero; a is a positive number; $-a$ is a positive number.

Axiom II — If both a and b are positive then both the sum $a + b$ and the product ab are also positive.

We will assume the truth of these axioms without formal proof, and infer from them the statement about binary relationships that for any two real numbers, a and b, one and only one of the following relationships must hold:

$$a = b \qquad a > b \text{ (or } b < a) \qquad a < b \text{ (or } b > a) \qquad (5\text{-}1)$$

The foregoing discussion deals only with strict inequalities, i.e., those inequalities that can be adequately expressed by symbols "$>$" and "$<$." Two other important relationships are represented symbolically as "$\geq$," which is translated as "is greater than or equal to," and "$\leq$," which is translated as "is less than or equal to." Consider the following inequalities.

$$a \geq b \qquad c \geq d \qquad e \leq f \qquad g \geq h$$

The "or" in the above translations is mutually exclusive according to the first axiom, that is, the symbol "$\geq$" means that the term to the left is greater than *or* equal to the one on the right, and conversely for the symbol $\leq$.

Operations with negative numbers are presented in some elementary books without a complete axiomatic explanation of the procedures involved, which are therefore reviewed here. Consider the following theorems.

Theorem 1.   If a positive number a is multiplied by a negative number b, the result is a negative number.

Theorem 2.   If a negative number a is multiplied by another negative number b, the result is a positive number.

In the case of the first theorem we may say that $-b$ is a positive number according to Axiom I. Then it follows from Axiom II that $a(-b)$ is a positive number. If we write the negative of this we obtain $-[a(-b)]$, which must be a negative number. But $-[a(-b)]$ is equal to ab. Consequently the product ab must be a negative number. This is an informal but adequate proof of Theorem 1. Theorem 2 may be proved by adopting a similar procedure. Then by combining Axiom II and Theorem 2 we may deduce that, for any real number $a^2 \geq 0$, if $a \neq 0$, then $a^2 > 0$.

60. Fundamental Operations.   This section presents fundamental operations that can be performed with inequalities. In place of formal proofs, examples are used to illustrate the validity of the conclusions.

Transitivity

If we have three real numbers a, b and c such that $a > b$ and $b > c$, then $a > c$.   (5-2)

$$7 > 2 \text{ and } 2 > 1 \quad \therefore \quad 7 > 1$$

Similarly, if we have three real numbers a, b and c such that $a < b$ and $b < c$, then $a < c$.   (5-3)

$$1 < 2 \text{ and } 2 < 7 \quad \therefore \quad 1 < 7$$

$$-7 < -2 \text{ and } -2 < -1 \quad \therefore \quad -7 < -1$$

Addition

If we have two inequalities, $a > b$ and $c > d$, then $a + c > b + d$.   (5-4)

$$9 > 5 \text{ and } 11 > 10 \quad \therefore \quad 9 + 11 > 5 + 10 \quad \text{or} \quad 20 > 15$$

$$-5 > -7 \text{ and } 5 > 4 \quad \therefore \quad -5 + 5 > -7 + 4 \quad \text{or} \quad 0 > -3$$

Similarly, if we have the single inequality $a > b$, and the real number c where $c \geq 0$, then $a + c > b + c$.   (5-5)

$$4 > 3 \text{ and } 8 > 0 \quad \therefore \quad 4 + 8 > 3 + 8 \quad \text{or} \quad 12 > 11$$

Likewise, if we have the two inequalities $a < b$ and $c < d$, then $a + c < b + d$.   (5-6)

$$0 < 5 \text{ and } 5 < 6 \quad \therefore \quad 5 + 0 < 5 + 6 \quad \text{or} \quad 5 < 11$$

Similarly if we have the single inequality $a < b$, and the real number c where $c \leq 0$, then $a + c < b + c$.   (5-7)

$$-3 < 0 \text{ and } 7 > 0 \quad \therefore \quad -3 + 7 < 0 + 7 \quad \text{or} \quad 4 < 7$$

Subtraction

If we have the two inequalities $a > b$ and $c > d$, then $a - d > b - c$.   (5-8)

$$9 > 7 \text{ and } 10 > 9 \quad \therefore \quad 9 - 9 > 7 - 10 \quad \text{or} \quad 0 > -3$$

$$7 > 6 \text{ and } 5 > 4 \quad \therefore \quad 7 - 4 > 6 - 5 \quad \text{or} \quad 3 > 1$$

Furthermore if we have the single inequality if $a > b$ and the real number c, where $c \geq 0$, then $a - c > b - c$.   (5-9)

$$7 > 6 \text{ and } 4 > 0 \quad \therefore \quad 7 - 4 > 6 - 4$$

If we have the two inequalities $a < b$ and $c < d$, then $a - d < b - c$.

$$(5\text{-}10)$$

$0 < 3$ and $-1 < 2$ $\therefore$ $0 - 2 < 3 - (-1)$ or $-2 < 4$

$-5 < -3$ and $3 < 5$ $\therefore$ $-5 - 5 < -3 - 3$ or $-10 < -6$

Similarly if we have the inequality $a < b$ and the real number c where $c > 0$, then $a - c < b - c$ $(5\text{-}11)$

$-5 < -3$ and $8 > 0$ $\therefore$ $-5 - 8 < -3 - 8$ or $-13 < -11$

However, given the two inequalities $a > b$ and $c > $ d, it is not justified to conclude that $a - c > b - d$.

Thus $5 > 2$ and $4 > 0$ but $5 - 4 \not> 2 - 0$ *

Moreover, given the two inequalities $a < b$ and $c < d$, it is not justified to conclude that $a - c < b - d$.

Thus $1 < 4$ and $0 < 4$ but $1 - 0 \not< 4 - 4$

Multiplication

If we have the inequality $a > b$ and the real number c where $c > 0$, then $ac > bc$. $(5\text{-}12)$

$5 > 3$ and $4 > 0$ $\therefore$ $(5)(4) > (3)(4)$ or $20 > 12$

If we have the inequality $a > b$ and the real number c where $c < 0$, then $ac < bc$ $(5\text{-}13)$

$5 > 3$ and $-4 < 0$ $\therefore$ $(5)(-4) < (3)(-4)$ or $-20 < -12$

If we have the inequality $a < b$ and the real number c where $c > 0$, then $ac > bc$. $(5\text{-}14)$

$2 < 7$ and $3 > 0$ $\therefore$ $(2)(3) < (7)(3)$ or $6 < 21$

If we have the inequality $a < b$ and the real number c where $c < 0$, then $ac > bc$. $(5\text{-}15)$

$2 < 7$ and $-3 < 0$ $\therefore$ $(2)(-3) > (7)(-3)$ or $-6 > -31$

*This sign, i.e. "$\not>$" should be translated as "is not larger than." Similarly any other equality sign or inequality sign that is cancelled in this way is said to be negated.

If we have the two inequality relations $a > b > 0$ and $c > d > 0$, then $ac > bd > 0$. (5-16)

$9 > 1 > 0$ and $5 > 4 < 0$ $\therefore$ $(9)(5) < (1)(4) < 0$ or $45 < 4 < 0$

If we have the two inequality relations $0 < a < b$ and $0 < c < d$, then $0 < ac < bd$ (5-17)

$0 < 2 < 3$ and $0 < 1 < 4$ $\therefore$ $0 < (2)(1) < (3)(4)$ or $0 < 2 < 12$.

Division

If we have the inequalities $a > b > 0$ and $c > d > 0$, then $a/d > b/c > 0$ (5-18)

$21 > 8 > 0$ and $4 > 3 > 0$ $\therefore$ $\dfrac{21}{3} > \dfrac{8}{4} > 0$ or $7 > 2 > 0$

However, if we have the two inequality relations, $a > b > 0$ and $c > d > 0$, it is not justified to conclude that $a/c > b/d$.

$$21 > 20 > 0 \text{ and } 7 > 1 > 0 \text{ but } \frac{21}{7} \not> \frac{20}{1}$$

If we have the inequalities $0 < a < b$ and $0 < c < d$, then $0 < \dfrac{a}{d} < \dfrac{b}{c}$. (5-19)

$$0 < 1 < 6 \text{ and } 0 < 3 < 4 \therefore 0 < \frac{1}{4} < \frac{6}{3}.$$

Exponents

If we have the inequality $a > b > 0$ and the real number p where $p > 0$, then $a^p > b^p$ and $a^{-p} < b^{-p}$. (5-20)

$$5 > 3 > 0 \text{ and } 2 > 0 \therefore 5^2 > 3^2 \text{ or } 25 > 9$$

and

$$5^{-2} < 3^{-2} \text{ or } \frac{1}{25} < \frac{1}{9}$$

If we have the inequality $a > b > 0$ and the real number p where $p < 0$, then $a^p < b^p$ and $a^{-p} > b^{-p}$ (5-21)

$$5 > 3 > 0 \text{ and } -2 < 0 \therefore 5^{-2} < 3^{-2} \text{ or } \frac{1}{25} < \frac{1}{9}$$

and

$$5^{-(-2)} > 3^{-(-2)} \text{ or } 5^2 > 3^2$$

61. Inequality of Averages. An important inequality is that existing between three widely used averages: the arithmetic mean, the geometric mean and the harmonic mean.

The arithmetic mean of a set of positive numbers may be obtained by dividing their sum by their number, e.g. the arithmetic mean of four numbers is their sum divided by four.

The harmonic mean of a set of positive numbers is obtained by dividing the number of the numbers by the sum of the reciprocals of the numbers.

The geometric mean of a set of n positive numbers is the nth root of their product.

The rule for inequality among these averages is that

$$\text{Arithmetic mean} > \text{Geometric mean} > \text{Harmonic mean}$$
$$(5\text{-}22)$$

Let a and b be two positive numbers, where $a \neq b$.
Then by the foregoing definitions,

$$\text{Arithmetic mean } (a, b) = M_A = \frac{a + b}{2}$$

$$\text{Harmonic mean } (a, b) = M_H = \frac{2}{\dfrac{1}{a} + \dfrac{1}{b}} = \frac{2ab}{a + b}$$

$$\text{Geometric mean } (a, b) = M_G = \sqrt{ab}$$

Then $\qquad M_A \cdot M_H = \left(\dfrac{a + b}{2}\right) \cdot \left(\dfrac{2ab}{a + b}\right) = ab = M_G^2$

or $\qquad M_A M_H = M_G^2$ so that $\dfrac{M_A}{M_G} = \dfrac{M_G}{M_H}$

Therefore the value of M_G is between those of M_A and M_H.

Furthermore $\qquad M_A - M_H = \dfrac{(a - b)^2}{2(a + b)} > 0$

Therefore $\qquad\qquad M_A > M_H.$

Application of the harmonic mean: A person travels from A to B with speed v_1 and returns with speed v_2. The average speed (defined as total distance travelled divided by the time necessary) is easily verified to be the harmonic mean of v_1 and v_2.

Example. Find the arithmetic, geometric and harmonic means of the numbers 60, 70, 80 and 90.

Since we are given no weighting factors, the arithmetic mean is simply the sum of the numbers divided by their number.

$$\text{Arithmetic mean} = \frac{60 + 70 + 80 + 90}{4} = 75.$$

The geometric mean is, by the statement above, the fourth root of the product of the four numbers

$$\text{Geometric Mean} = \sqrt[4]{60 \cdot 70 \cdot 80 \cdot 90}$$
$$= \sqrt[4]{30,240,000}$$
$$= 74.155$$

The harmonic mean, since there are no weighting factors, is the number of numbers divided by the sum of their reciprocals.

$$\text{Harmonic Mean} = \frac{4}{\dfrac{1}{60} + \dfrac{1}{70} + \dfrac{1}{80} + \dfrac{1}{90}}$$

$$= \frac{4}{\dfrac{84}{5040} + \dfrac{72}{5040} + \dfrac{63}{5040} + \dfrac{54}{5040}}$$

$$= \frac{4}{\dfrac{275}{5040}} = \frac{20,160}{275}$$

$$= 73.309$$

Thus we have $75 > 74.155 > 73.309$

or arithmetic mean > geometric mean > harmonic mean.

62. Proper-Name Inequalities. There are a number of inequalities in mathematics which are known by proper names, just as there are other mathematical relations that are so named. One of the best known of the proper-name inequalities is that of Schwarz. It applies to vectors, for example, where it takes the form (denoting the inner vector product by (a, b) and the corresponding matrices by $\| a \|$ and $\| b \|$.

$$| (a, b) |^2 \leq \| a \|^2 \cdot \| b \|^2 \tag{5-23}$$

It also applies to complex numbers in the form

$$| a_1\bar{b}_1 + a_2\bar{b}_2 + \cdots + a_n\bar{b}_n |^2 \leq | a_1\bar{a}_1 + a_2\bar{a}_2 + \cdots + a_n\bar{a}_n | \cdot$$
$$| b_1\bar{b}_1 + b_2\bar{b}_2 + \cdots + b_n\bar{b}_n | \quad (5\text{-}24)$$

while a third type of mathematical object to which it applies are the integrals of mathematical functions. Here it takes the form

$$\left| \int_a^b \bar{f}_1(x)f_2(x)dx \right|^2 \leq \int_a^b \bar{f}_1(x)f_1(x)dx \cdot \left| \int_a^b \bar{f}_2(x)f_2(x)dx \right|$$
$$(5\text{-}25)$$

where $f_1(x)$ and $f_2(x)$ are two functions of x, and $\bar{f}_1(x)$ and $\bar{f}_2(x)$ are their complex conjugates. (Complex conjugates are related to their corresponding complex numbers or functions in that the sign of the imaginary part is reversed. Thus $a + ib$ and $a - ib$ are conjugate complex numbers.)

63. Systems of Inequalities. Before considering the significance of a system of inequalities, let us first pause briefly to review the graphical solution of a system of linear equations. In Figure 5-2 below, two linear equations in two unknowns are plotted on the same graph. The plotting is effected by selecting a value for x, and then solving the equation for y. When this operation has been performed twice for one of the equations, a line is drawn between the two points that have been determined. By finding other points in a similar manner, the line may be extended indefinitely in either direction. Ths procedure is then applied to the second equation to obtain its graph. Unless the two lines representing these equations are parallel, there is a common point of intersection and this point indicates the values of x and y that will satisfy both equations simultaneously. In the case illustrated below graphs #1 and #2 intersect at the point (2, 4.5). These values are the required solution for this set of linear equations.

Thus we can say that if a set of linear equations has a common solution, it is represented graphically as a point determined by the intersection of two or more lines. (See discussion in Chapter 2 of Rank of a Matrix and Consistency and Independence of Equations.)

Systems of linear inequalities may also be defined graphically but the solution of such a system cannot usually be represented as a single point. Quite often the solution may be all the points on a given line or all the points on a specific part of a plane.

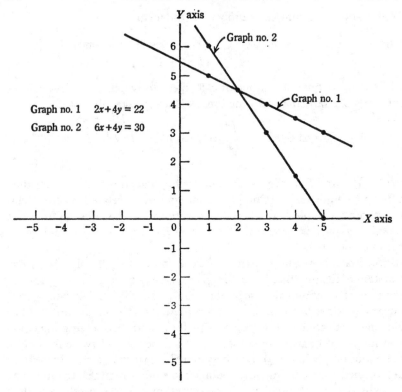

FIG. 5-2. Graphical solution of two first degree simultaneous equations in two variables.

For example, in Figure 5-3 next page, the line segment designated by the numeral I represents the following set of inequalities:

$$x \geq -10$$

$$x \leq +10$$

$$y = +5$$

Any point lying on the line between the points $(-10, 5)$ and $(+10, 5)$ provides an acceptable solution to this set of inequalities.

The numeral II in Figure 5-3 designates a ray beginning at the point $(0, -8)$ and extending indefinitely to the left of the Y-axis in a direction perpendicular to the Y-axis. This ray represents the expressions

$X \leq 0$ and $y = -8$. Any point lying on ray II above is a solution to this system.

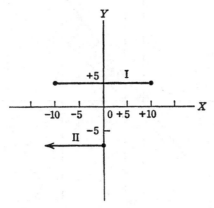

FIG. 5-3.   The graphical representation
of two sets of inequalities.

Both of these examples are quite elementary.   Let us now consider a slightly more complex system.   In Figure 5-4 we have the graphical representation of the following system of inequalities:

$$\text{I. } y \leq \tfrac{4}{3} x + 4$$

$$\text{II. } x \geq -3$$

$$\text{III. } y \geq x$$

$$\text{IV. } x \leq +10$$

If each of the inequality signs in the above expressions were replaced by an equality sign, the four equations could be represented by the four straight lines that bound the shaded quadrilateral in Figure 5-4. However, the solution to this system of inequalities is the set of all points lying within or on the boundary lines of the shaded quadrilateral.   If we regard inequality I above as the sum of an equation $y = \tfrac{4}{3} x + 4$, and an inequality $y < \tfrac{4}{3} x + 4$, then the former represents all points lying on line I in Figure 5-4 and the latter, all points in the XY plane lying below that line.   In other words, the line plus the designated portion of the XY plane satisfies inequality I which states that "y is less than *or* equal to $\tfrac{4}{3} + 4$."

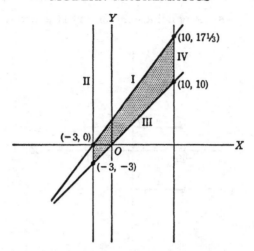

FIG. 5-4   The graphical representation of a quadrilateral as defined
by a system of inequalities.

Similarly the inequality $x \geq -3$ is represented by the line $x = -3$
(II in the figure) and all points in the XY plane lying to the right of this
line.   The inequality $y \geq x$ is represented by the line $y = x$ (III in the
figure) and all points in the XY plane lying above this line.   Finally,
the inequality $x \leq 10$ is represented by the line $x = 10$ (IV in the
figure) and all the points in the XY plane lying to the left of this line.
Those points that satisfy all four of these inequalities are those described
above, i.e., those points lying within and on the boundaries of the
shaded quadrilateral.

64. Linear Programming.   There are many practical problems
that involve the maximization or minimization of a linear function over
a corresponding set of possible values for the given variables.   Many of
these problems are solved by the methods of linear programming.

Let us return to the system of inequalities represented in the preceding section by Figure 5-4.   We had

$$\text{I. } y \leq \tfrac{4}{3} x + 4$$

$$\text{II. } x \geq -3$$

$$\text{III. } y \geq x$$

$$\text{IV. } x \leq +10$$

Now consider that x and y are two types of product and that a business may market them subject to the restrictions of the above inequalities. Assume further that the profit on x is 2 units and that on y is 3 units. That is, the two products may be sold in any combination that satisfies the above inequalities. For example the business could not sell 11 of x and 2 of y, for the point $(11, 2)$ does not satisfy our system of inequalities. However, it could sell a combination of 8 of x and 10 of y, for this point does satisfy the system. Furthermore, the business makes a profit of $2x + 3y$ for any sale of items x and y. What sales combination then provides the greatest profit for the business?

We know that the combination must be represented by a point or points lying on the boundaries or within the quadrilateral of Figure 5-4. Let us first find the values for $2x + 3y$ obtained by substituting the coordinates of the corner points for x and y.

TABLE 5-1

x	y	$2x + 3y$
10	$17\frac{1}{3}$	72
10	10	50
−3	0	−6
−3	−3	−15

In Table 5-1 we see that the corner point defined by the coordinates $(10, 17\frac{1}{3})$ yields the highest profit of any of these four points. Upon further inspection we find that no other available combination will yield a greater profit. Similarly the point $(-3, -3)$ indicates the greatest loss possible under the given conditions. (The negative values of x and y represent returns of previously sold goods for credit.)

This example illustrates a very important relationship between our study of inequalities and a linear equation of the form $ax + by + c$. If we have a system of inequalities such that its graphical representation is a polygon of finite area and if we have a related function of the form $ax + by + c$, both the maximum and the minimum value of the function subject to the given constraints occur at specific corner points of the polygon. Consequently, the values of the linear function, found by taking the coordinates of points within the polygon or of points lying on the border of the polygon (excluding corner points), always lie between a maximum and a minimum defined by a pair of corner points. Let us consider another example of a linear programming problem.

A factory has been given the opportunity of displaying 20 items at an industrial fair. This factory manufactures three distinct products, all

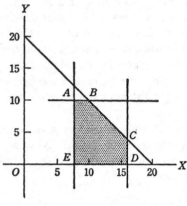

FIG. 5-5.

of which come in numerous models. We designate the three products as x, y, and z (In the calculations $20 - x - y$ will replace z). Since product x is very popular, the president of the company demands that between 8 and 16 of these be sent to the fair. Product y is less popular and the president's only restriction here is that 10 or less of these be sent. Product z is new and the president places no restrictions on it. Thus we have the following system of inequalities which is illustrated in Figure 5-5.

$$x + y \leq 20$$

$$x \leq 16$$

$$x \geq 8$$

$$y \leq 10$$

$$y \geq 0$$

In the above figure, the shaded polygon and its boundaries represent the set of points that are possible solutions to the given system of inequalities. It is also known that the cost of shipping the products is represented by the equation:

$$C = 3x + 2y + (20 - x - y)$$

In other words, the shipping expenses for x, y, and z are in the ratio $3 : 2 : 1$. Furthermore, the president of the company believes that each sample of product x will attract 10 times as many buyers as one sample of product y and that each sample of product z will attract 4 times as many buyers as one sample of product y. Thus, we have the equation

$$B = 10x + y + 4(20 - x - y)$$

The president of the company is interested in knowing how many of each product should be sent to meet either of two considerations. They are: first, minimization of the shipping cost; and second, maximization of buyer appeal. To discover the numbers of products yielding these extreme values, it is necessary to take the coordinates of each of the corner points in the above figure and introduce these numbers into the two linear equations for cost and buyer appeal. These results are given in Table 5-2 below.

TABLE 5-2

Point	x	y	z*	Cost of Shipping $3x + 2y + (20 - x - y)$	Buyer Appeal $10x + y + 4(20 - x - y)$
A	8	10	2	46	98
B	10	10	0	50	110
C	16	4	0	56	164
D	16	0	4	52	176
E	8	0	12	36	128

*$z = (20 - x - y)$

From the results in the above table, we conclude that, if shipping cost is no object, the factory should send 16 of product x and 4 of product z for maximum buyer appeal. On the other hand, if it is necessary to operate at minimum cost, the factor should send 8 of product x and 12 of product z to the fair. With these figures available, a compromise decision may be reached.

However, if the president now wishes to include at least one of product y, we add the inequality $y \geq 1$ to our system and disregard the inequality $y \geq 0$. This change is shown in Figure 5-6.

In the new polygon points A, B, and C have the same coordinates as in the original polygon, but point D has become D' and point E has become E'. The cost and buyer appeal of these new corner points is shown in the Table 5-3.

TABLE 5-3

Point	x	y	z*	Cost of Shipping $3x + 2y + (20 - x - y)$	Buyer Appeal $10x + y + 4(20 - x - y)$
D'	16	1	3	53	173
E'	8	1	11	37	125

*$z = 20 - x - y$

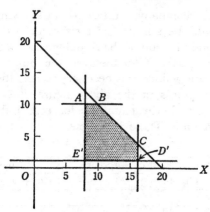

Fig. 5-6.

Under this new condition the factory should send 16 of x, 1 of y, and 3 of z for maximum buyer appeal or 8 of x, 1 of y, and 11 of z for minimum cost.

By following the foregoing procedure, many elementary problems in linear programming may be solved with comparitive ease. The five major steps in this procedure are outlined below.

1. Translate the limiting conditions into an equivalent system of inequalities.

2. Find the corner points of the polygon defined by the system of inequalities.

3. Set up a linear equation for the quantity that is to be maximized or minimized.

4. Substitute the coordinates of the corner points into the linear equation and tabulate the results.

5. Look for the required value of the linear function (i.e., the maximum or the minimum value).

65. The Transportation Problem. In the preceding section we were concerned with problems which enta.led a small number of variables and which, consequently, could be solved graphically. The problems of business and industry can rarely be solved with such ease.

In this section we will consider a more complex problem involving a greater number of variables. Even so, the example used here is still very elementary, though it is more than adequate for the demonstration of a technique which can be applied to more advanced problems.

One of the fundamental concerns of industry is to minimize the cost of shipping materials from one place to another.   The "industrial fair" problem of the preceding section is one example of this type.   However, there is a great multiplicity of such problems.

Let us now consider a situation in which the products of 3 factories are to be assigned to 4 warehouses.   The object, of course, is to minimize the cost of shipping the products from one place to another. Factories will be denoted by F_1, F_2, and F_3: warehouses by W_1, W_2, W_3, and W_4.   Tables 5-4 and 5-5 indicate the material (in tons) that must be shipped from the factories and the capacity (in tons) of the respective warehouses.

TABLE 5-4

Factory	F_1	F_2	F_3
Material to be Shipped	5	60	40

TABLE 5-5

Warehouse	W_1	W_2	W_3	W_4
Capacity	35	10	35	25

We see here that there is a total of 105 tons that must be removed from the three factories and that the total capacity of the four warehouses is also 105 tons.   This is a special situation, for it is not usually the case that these two totals are equal.   Under these conditions the problem may be solved by using a system of equalities rather than a system of inequalities.   Although this problem may seem oversimplified, it provides a good example for illustrating the basic techniques of solution.   After we have solved this first problem we will revise it slightly so that the equalities become inequalities and the situation will then assume a more general character.

Thus, in this case, if we let f_i $\cdots$ represent the amounts that must be shipped from the factories and let w_j $\cdots$ represent the capacities of the warehouses, we have the following equation:

$$\sum_{i=1}^{3} f_i = \sum_{j=1}^{4} w_j \tag{5-26}$$

Furthermore, if we let x_{ij} denote the amount that is to be shipped from the ith factory to the jth warehouse, we may also set up the following equations

$$\sum_{j=1}^{4} x_{ij} = f_i \text{ for each factory} \tag{5-27}$$

$$\sum_{i=1}^{3} x_{ij} = w_j \text{ for each warehouse.} \tag{5-28}$$

Equation (5-27) simply states that each factory must ship the amount indicated in table 5-4. Likewise equation (5-28) states that each warehouse must be assigned its capacity as indicated by Table 5-5. A series of equations may be written from the information contained in Tables 5-4 and 5-5 and equations (5-26), (5-27) and (5-28). We have

$$\left.\begin{array}{l} x_{11} + x_{12} + x_{13} + x_{14} = 5 \\ x_{21} + x_{22} + x_{23} + x_{24} = 60 \\ x_{31} + x_{32} + x_{33} + x_{34} = 40 \end{array}\right\} \text{out of factories} \qquad (5\text{-}29)$$

and

$$\left.\begin{array}{l} x_{11} + x_{21} + x_{31} = 35 \\ x_{12} + x_{22} + x_{32} = 10 \\ x_{13} + x_{23} + x_{33} = 35 \\ x_{14} + x_{24} + x_{34} = 25 \end{array}\right\} \text{into warehouses} \qquad (5\text{-}30)$$

The other important factor in this problem is the cost of transportation. In Table 5-6 below the freight rates are given in dollars per ton for shipping from any factory to any warehouse.

TABLE 5-6

		Warehouses			
		W_1	W_2	W_3	W_4
Factories	F_1	$1.05	$2.30	$1.80	$1.00
	F_2	$.90	$1.40	$1.00	$1.75
	F_3	$2.00	$1.40	$1.20	$1.10

Thus to ship one ton from Factory 2 to Warehouse 3 costs $1.00, etc. Therefore, the total cost of the operation is:

$$1.05\, x_{11} + 2.30\, x_{12} + 1.80\, x_{13} + 1.00\, x_{14} + \cdots + 1.10\, x_{34} = Cost = C$$

or $C = \displaystyle\sum_{i=1}^{3}\sum_{j=1}^{4} c_{ij}x_{ij} \geq 0$ \qquad (5-31)

where c_{ij} is the cost of shipping one ton of material from factory i to warehouse j. This total cost $C = \displaystyle\sum_{i=1}^{3}\sum_{j=1}^{4} c_{ij}x_{ij}$ is to be minimized. We may now combine all this information into one master table (Table 5-7).

TABLE 5-7

		Warehouses				Factory
		W_1	W_2	W_3	W_4	Excess
		Freight rates				
Factories	F_1	$1.05	$2.30	$1.80	$1.00	5
	F_2	$.90	$1.40	$1.00	$1.75	60
	F_3	$2.00	$1.40	$1.20	$1.10	40
Warehouse Capacities		35	10	35	25	105 Total

In general practice it is possible to make an educated guess at what might seem to be a good solution, and this tends to shorten the search for an optimal solution.   However, it is not always possible to operate on this kind of estimate and we will treat this problem on an arbitrary assignment basis so that the general procedure will be clear for more difficult problems.

That is, we begin by arbitrarily allocating the factory excess to the warehouses according to Table 5-8.

TABLE 5-8

		Warehouses				Totals
		W_1	W_2	W_3	W_4	
	F_1	5				5
Factories	F_2	30	10	20		60
	F_3			15	25	40
Totals		35	10	35	25	105

These allocations were made by: (1) assigning all of F_1's excess to W_1. (2) Filling W_1's capacity with 30 tons from F_2. (3) Filling W_2's capacity with 10 tons from F_2. (4) Assigning the balance of F_2's excess to W_3. (5) Filling W_3's capacity with 15 tons from F_3. (6) Assigning the balance of F_3's excess to W_4, filling its capacity. This method makes no attempt to minimize the shipping costs but merely provides us with a starting point for our search for the optimal solution.

The next step is to create a table of rates based on the allocation of materials according to Table 5-8.   This is accomplished by first inserting the actual freight rates, taken from Table 5-6 or Table 5-7, for those routes which have been designated for use.   (See Table 5-9 below).

Then fill in the "row values" and "column values."   To do this assign an arbitrary row value to Row F_1; we have chosen .00 for this value, but it could have been any real number.   Once this first value has been assigned, it is possible to obtain a complete set of row and column values provided that the rate table is not "degenerate." The notion of degeneracy will be elaborated on later in this chapter. Since the tables involved in the expression of this problem have three rows and four columns it becomes necessary to determine three row and four column values. The row and column values are constructed so that each rate corresponding to a proposed allocation can be expressed as the sum of its corresponding row and column value. Since a value of .00 is assigned to row F_1, we are compelled to assign the value 1.05 to column W_1 in order that the sum of these two values equal the rate at the intersection F_1W_1, i.e., 1.05.

Then, since W_1 has been assigned the value 1.05, the row value for F_2 can be determined, for we know that the row value for F_2 plus the column value for W_1 (1.05) must equal .90, the value at the appropriate intersection (F_2W_1). Thus the row value for F_2 is found to be $-.15$, since $-.15 + 1.05 = .90$. Using this figure $(-.15)$ in conjunction with the rates appearing in row F_2 and columns W_2 and W_3, it is then possible to find column values for the latter. The column value of W_2 when added to $-.15$ must yield 1.40, and is therefore 1.55. Similarly the column value of W_3 when added to $-.15$ must yield 1.00, and is thus 1.15. Thus we obtain 1.55 and 1.15 respectively for the W_2 and W_3 column values. Continuing this process we know that the column value for $W_3(1.15)$ plus the row value for F_3 must equal 1.20, and consequently we obtain .05 as the desired row value for F_3. To complete the assignment of row and column values we find the column value for W_4 by using the row value $F_3(.05)$ and 1.10, the rate at the corresponding intersection (F_3W_4). Thus the column value for W_4 is found to be 1.05.

TABLE 5-9

		Warehouses				Row
		W_1	W_2	W_3	W_4	Values
	F_1	$1.05				.00
Factories	F_2	$.90	$1.40	$1.00		−.15
	F_3			$1.20	$1.10	.05
Column Values		1.05	1.55	1.15	1.05	

The construction of Table 5-9 shows that where there are p rows and q columns and $p + q - 1$ entries in the rate table corresponding to a proposed system of allocations, the entire set of row and column values for this table are uniquely determined as soon as one value has been arbitrarily assigned.

We now devise a cost table based on the allocations made in Table 5-8. This is illustrated by Table 5-10, in which the row values and column values have been taken from Table 5-9, and the numbers in boldface type represent the actual cost values also as extracted from Table 5-9, while the numbers in italic type represent artificial cost values, derived by adding an appropriate row value to an appropriate column value to give a sum which is inserted at the corresponding intersection of the selected row and column.

TABLE 5-10

		Warehouses				Row Values
		W_1	W_2	W_3	W_4	
Factories	F_1	**$1.05**	*$1.55*	*$1.15*	*$1.05*	.00
	F_2	**$.90**	**$1.40**	**$1.00**	*$.90*	−.15
	F_3	*$1.10*	*$1.60*	**$1.20**	**$1.10**	.05
Column Values		1.05	1.55	1.15	1.05	

For example, row 1 was completed by adding the respective column values to .00, the row value of the first row, and inserting the results at the appropriate intersections.

Now we are ready to revise our original allocations and to proceed in our search for the optimal solution. First we compare the artificial costs in Table 5-10 with the actual costs in Table 5-7, looking for the figure in Table 5-10 that exceeds by the greatest amount the corresponding figure in Table 5-7. We find that there are only two artificial costs that are greater than their corresponding actual costs. These occur at the intersection of row F_3 and column W_2 and at the intersection of row F_1 and column W_4. Of these the former 1.60 (Table 5-10) exceeds its actual value 1.40 (Table 5-7) by .20 and the latter 1.05 (Table 5-10) exceeds its actual value 1.00 (Table 5-7) by .05. Thus the intersection at F_3 and W_2 is the greatest excess, hence the required figure. This difference of .20 tells us that if we make shipments from F to W_2, and make the necessary adjustments in the rest of our program, we shall save 20 cents for every ton shipped along this new route.

In order to discover what adjustments will be necessary in our original program if we plan to introduce this new route, we construct

Table 5-11 by first reproducing Table 5-8 and then adopting the following procedure.

TABLE 5-11

		Warehouses				
		W_1	W_2	W_3	W_4	Totals
Factories	F_1	⑤				5
	F_2	㉚	$10 - \theta$	$20 + \theta$		60
	F_3		$+\theta$	$15 - \theta$	㉕	40
Totals		35	10	35	25	1.05

At the intersection of F_3 and W_2 we write $+\theta$ which represents the, as yet, undetermined amount that will be shipped on this new route. This addition has overloaded W_2 by the amount θ. Consequently we must balance the shipment to this warehouse by subtracting θ from the entry at the intersection of F_2 and W_2. Now F_2 has been depleted and to bring its total back to normal it is necessary to add θ to the entry at the intersection of F_2 and W_3. This change allocates too much material to W_3; thus, θ must now be subtracted from the entry at the intersection of F_3 and W_3. This last change also balances the F_3 row which became overloaded upon our original insection of θ. Once the unknown quantity θ has been introduced into the table, it is necessary to balance all of the affected totals by this procedure of alternately adding and subtracting θ to or from some of the entries. To determine what entries are not affected by the introduction of θ, we first write θ at the designated intersection and regard it as a real number. Then we circle every entry in the table that is the only number in *either* its row *or* its column. The entries at the intersections of F_1 and W_1 and of F_3 and W_4 may be circled and consequently disregarded in the balancing procedure. If we now consider these two entries to be eliminated, we repeat the first operation and find that the entry at the intersection of F_2 and W_1 is now in a column by itself and may be similarly eliminated. Continuing this procedure we find that none of the remaining entries are alone in either a column or a row. Thus we have made all the possible eliminations. Then we proceed to balance the table by systematically subtracting and adding θ to the remaining entries. Our progress at this stage is illustrated above in Table 5-11.

Since every ton shipped along the new route means a savings of 20 cents, it is obviously desirable to divert as much tonnage as possible to this route. Therefore, we examine all the entries in Table 5-11 that contain a $-\theta$ term.   The smallest number followed by $-\theta$ is the 10 at the intersection of F_2 and W_2.   This is the limit of the amount that can be diverted and consequently is also the required value of θ.   If θ was chosen to be more than 10, we would obtain a negative entry at the intersection F_2W_2 and this is clearly impossible.   At this point we obtain Table 5-12 by reproducing Table 5-11 and performing the indicated additions and subtractions with θ equal to 10.

TABLE 5-12

		Warehouses				
		W_1	W_2	W_3	W_4	Totals
Factories	F_1	5				5
	F_2	30		30		60
	F_3		10	5	25	40
Totals		35	10	35	25	105

To justify what we have done so far, let us compare the total cost of our original set of allocations with the total cost of our first revised set of allocations.   This is accomplished by evaluating the total cost for each, using equation (5-31), $C = \sum\limits_{i=1}^{3} \sum\limits_{j=1}^{4} c_{ij}x_{ij}$.   All terms c_{ij} are found in Table 5-6, our original table of costs.   The x_{ij} for our original set of allocations are found in Table 5-8 and those for our first revised set of allocations are found in Table 5-12.   In tables 5-8 and 5-12, the values of all x_{ij} at intersections not having a numerical entry is zero.   Thus let C_0 indicate the cost according to the original allocations and let C_1 indicate the cost according to our revised allocations.   We then have:

$$C_0 = \$1.05(5) + \$.90(30) + \$1.40(10) + \$1.00(20) + \$1.20(15)$$
$$+ \$1.10(25)$$

$$= \$5.25 + \$27.00 + \$14.00 + \$20.00 + \$18.00 + \$27.50$$

$$= \$111.75$$

$$C_1 = \$1.05(5) + \$.90(30) + \$1.00(30) + \$1.40(10) + \$1.20(5)$$
$$+ \$1.10(25)$$

$$= \$5.25 + \$27.00 + \$30.00 + \$14.00 + \$6.00 + \$27.50$$

$$= \$109.75$$

$$C_0 - C_1 = \$111.75 - \$109.75 = 2.00$$

This difference of $2.00 is equal to the 20¢ per ton difference in rate multiplied by the 10 tons that were rerouted. Thus we have made an improvement.

To improve our allocations still more we must duplicate the procedure that we used on the original table of allocations. We proceed as follows.

First, we create a new cost table (Table 5-13) by copying from Table 5-6 the actual rates for the routes used in Table 5-12. These rates are again shown in boldface type. Then the row and column values are determined and the artificial costs are filled in to complete the table (italic type).

TABLE 5-13

		Warehouses				Row
		W_1	W_2	W_3	W_4	Values
Factories	F_1	**1.05**	*1.35*	*1.15*	*1.05*	.00
	F_2	**.90**	*1.20*	**1.00**	**.90**	−.15
	F_3	*1.10*	**1.40**	**1.20**	**1.10**	.05
Column Values		1.05	1.35	1.15	1.05	

Next we examine the artificial costs in Table 5-13 to see if any are greater than the actual costs in Table 5-6. We find only one such entry: at the intersection of F_1 and W_4 the artificial cost is $1.05 and the actual cost is $1.00. Thus we know that it is possible to save an additional .05 per ton that we can transfer to this route.

Now we set up Table 5-14 by duplicating Table 5-12, inserting θ at the new route, and by eliminating (circling) any element which stands alone *either* in its row or in its column. Only the element 10 at the

TABLE 5-14

		Warehouses				
		W_1	W_2	W_3	W_4	Totals
Factories	F_1	$5 - \theta$			θ	5
	F_2	$30 + \theta$		$30 - \theta$		60
	F_3		⑩	$5 + \theta$	$25 - \theta$	40
Totals		35	10	35	25	105

intersection of F_3 and W_2 may be eliminated in this fashion. Now we proceed systematically to add and subtract θ to the remaining entries in the table.   Of the terms which contain $-\theta$, the entry at the intersection of F_1 and W_1 has the smallest number followed by $-\theta$.   This number is 5, the limit of diversion, and consequently the value of θ.   Thus we can prepare Table 5-15 by substituting in Table 5-14 a value of 5 for θ wherever it occurs.

TABLE 5-15

		Warehouses				
		W_1	W_2	W_3	W_4	Totals
	F_1				5	5
Factories	F_2	35		25		60
	F_3		10	10	20	40
Totals		35	10	35	25	105

To find the total cost using this set of allocations we again employ equation (5-31),

$$C = \sum_{i=1}^{3} \sum_{j=1}^{4} c_{ij} x_{ij}$$

and we designate the new total cost as C_2,

$C_2 = \$1.00(5) + \$.90(35) + \$1.00(25) + \$1.40(10) + \$1.20(10)$
$\qquad + \$1.10(20)$

$C_2 = \$5.00 + \$31.50 + \$25.00 + \$14.00 + \$12.00 + \$22.00$

$C_2 = \$109.50$

$C_1 - C_2 = \$109.75 - \$109.50 = \$.25$

This difference of .25 is equal to the 5¢ per ton difference in rate multiplied by the 5 tons that were rerouted.

If we continue this procedure and set up a new table of actual and artificial costs based on the allocations of Table 5-15, we find that none of the artificial costs are greater than the actual costs and thus we have arrived at the optimal solution.   If it had been the case that one of the artificial values was equal to its corresponding actual value, this would have indicated that there was an alternative optimal solution which had the same cost (i.e., \$109.50) but with a different allocation of goods. As it is, there are no possible improvements that can be made upon the allocations of Table 5-15.

66. The Transportation Problem Revised.   Now we will make one small but very significant modification in the transportation problem of the foregoing section.   This change will at first produce a system of inequalities, which is more typical of the form of a transportation problem than the system of equalities in the preceding section.

For this problem we will change the excess at factor #1 from 5 tons to 15 tons and make no adjustment in the capacities of the various warehouses.   This means that one or more of the factories will have to store part of their excess at the factory itself.

However, although both factory #2 and factory #3 are able to store up to 10 tons, factory #1 has no provision for storage and consequently must ship its full 15 tons.   Thus we now have:

$$\sum_{i=1}^{3} f_i = \sum_{j=1}^{4} w_j + 10 \qquad (5\text{-}32)$$

where f_i is the excess at the "ith" factory and w_j is the capacity of the "jth" warehouse.   If x_{ij} denotes the amount that is to be shipped from the ith factory to the jth warehouse, we obtain the following expressions.

$$\sum_{i=2}^{3} \sum_{j=1}^{4} x_{ij} \leq f_i \text{ for each factory} f_i, \text{ where } i = 2 \text{ or } 3. \qquad (5\text{-}33)$$

and

$$\sum_{j=1}^{4} x_{1j} = f_1 \qquad \qquad 5\text{-}34)$$

$$\sum_{i=1}^{3} x_{ij} = w_j \text{ for each warehouse.} \qquad (5\text{-}35)$$

The relations (5-33) and (5-34) indicate that the amount shipped from $F_2 + F_3$ is less than or equal to their excess but that the amount shipped from F_1 is equal to its excess.   Equation (5-35) simply indicates that all warehouses must be filled.   From this information we see that we are dealing with a series of expressions of the following form.

$$\left.\begin{array}{l} x_{11} + x_{12} + x_{13} + x_{14} = 15 \\[4pt] x_{21} + x_{22} + x_{23} + x_{24} \leq 60 \\[4pt] x_{31} + x_{32} + x_{33} + x_{34} \leq 40 \end{array}\right\} \quad \text{out of factories} \qquad (5\text{-}36)$$

and

$$x_{11} + x_{21} + x_{31} = 35$$
$$x_{12} + x_{22} + x_{32} = 10$$
$$x_{13} + x_{23} + x_{33} = 35$$
$$x_{14} + x_{24} + x_{34} = 25$$

into warehouses.   (5-37)

The best way to attack this problem is to use a "dummy variable" Z, which will represent, in reality, the amount that must be stored at the factories.   However, in our tables we will consider Z to be a "dummy" warehouse with a capacity of 10 tons.   The cost of transporting material from F_1 to Z *will* be designated by M, which we will consider to be a very large number, since factory F, has no storage capacity. The cost of transporting material from F_2 to Z or from F_3 to Z will be designated as zero, since it is possible to store up to 10 tons at either of these locations.   By placing a large value at the intersection of F_1 and Z, we have made certain that no material will be allocated to this route in our optimal solution.   All of this information is illustrated below in Table 5-16.

TABLE 5-16

		Warehouses					Factory
		W_1	W_2	W_3	W_4	Z	Excess
		Freight Rates					
	F_1	1.05	2.30	1.80	1.00	M	15
Factories	F_2	.90	1.40	1.00	1.75	0	60
	F_3	2.00	1.40	1.20	1.10	0	40
Warehouse Capacity		35	10	35	25	10	

We now proceed exactly as we did in the preceding example.   First allocate the material.   Second set up a table of actual and artificial costs by determining the row and column values.   If any of the artificial costs are higher than their corresponding actual costs, it indicates that some route changes will be necessary.   If such cases exist we select the entry whose artificial cost exceeds its actual cost by the greatest amount and write θ into that square in our first route table.   Then the table is balanced and θ is found, the result being a revised route table This procedure is continued until we obtain a cost table in which no artificial cost exceeds the actual cost as defined in Table 5-16. The allocations proposed by this table provide the optimal solution of the problem. The step-by-step solution of this problem is left for the reader. The ultimate optimal allocations are given below in Table 5-17.

TABLE 5-17

		Warehouses					Totals
		W_1	W_2	W_3	W_4	ζ	
Factories	F_1			15			15
	F_2	35		25			60
	F_3		10	10	10	10	40
Totals		35	10	35	25	10	115

Let us check this solution by examining the corresponding cost table (Table 5-18 below).

TABLE 5-18

		Warehouses					Row Values
		W_1	W_2	W_3	W_4	ζ	
Factories	F_1	*1.00*	*1.30*	*1.10*	1.00	*−.10*	.00
	F_2	.90	*1.20*	1.00	.90	*−.20*	−.10
	F_3	*1.10*	1.40	1.20	1.10	0	.10
Column Values		1.00	1.30	1.10	1.00	−.10	

Upon comparing the artificial costs in Table 5-18 (those in italic type) with the actual costs in Table 5-16, we find that there is no artificial cost which exceeds an actual cost. Consequently Table 5-17 does represent an optimal solution to the problem.

The minimum cost, obtained by using this set of allocations is found by equation (5-31) to be:

$$C = \sum_{i=1}^{3} \sum_{j=1}^{4} c_{ij}x_{ij} = \$.90(35) + \$1.40(10) + \$1.00(25) + \$1.20(10)$$
$$+ \$1.00(15) + \$1.10(10) + \$.00(10)$$
$$= \$31.50 + \$14.00 + \$25.00 + \$12.00 + \$15.00$$
$$+ \$11.00 = \$108.50$$

67. Degeneracy and Formal Characteristics. The procedure illustrated above may be applied to transportation problems of any size provided that degeneracy does not appear in a route table at some stage of the solution.   A route table is said to be degenerate if it can be partitioned into two or more sections, each of which contains a group of destinations whose combined capacity exactly satisfies the combined requirements of the sources assigned to them.   Under these conditions it is impossible to construct a cost table corresponding to the degenerate route table.   Consider Table 5-19 below.   This is a degenerate route table but it is also a possible solution to the first transportation problem described above.

TABLE 5-19

		Warehouses				Totals
		W_1	W_2	W_3	W_4	
Factories	F_1		5			5
	F_2	35			25	60
	F_3		5	35		40
Totals		35	10	35	25	

In this table the excess from F_2 exactly exhausts the combined capacity of W_1 and W_4.   Likewise, the excess from F_1 and F_3 equals the combined capacity of $W_2 + W_3$.

Here is one simple technique that may be used to cope with degeneracy.   If there are fewer factories than warehouses, we divide one unit of shipment (smallest unit of shipment, e.g., $1/10$ ton) by twice the number of factories.   Then we take any convenient number that is smaller than this quotient and add it to the excess at each of the factories; then add the same total amount to any one of the warehouses. This procedure is reversed if the number of warehouses is smaller than its number of factories.

The problem may now be solved by following the usual procedure and regarding the additional quantities as real parts of the requirements and capacities.   When the problem has been solved, all numbers are rounded to the nearest unit of shipment.   A route carrying less than one half unit is rounded to zero.   The solution obtained by following this technique is not an approximation but is exact.

A great variety of linear programming problems may be solved by using the method of the transportation problem.   There is no great difficulty in changing the goal of the problem from minimization of cost to maximization of profit.   In the latter case a "margin table" replaces the cost table, the margin being the proceeds from selling a unit less the variable costs of producing it.   The same procedure is used to find the optimal solution except that new "routes" are introduced when the "artificial margin" is less than the actual margin.

To be solved by the transportation procedure, a problem should meet the following set of conditions. First, one unit of any input may be used to produce one unit of any output. Second, the cost or margin resulting from the conversion of one unit of a particular input into one unit of a particular output can be expressed by a single figure regardless of the number of units converted. Third, the quantity of each individual input and output is fixed in advance, and the total of the inputs equals the total of the outputs.

There are, however, a number of quasi-mathematical expedients which may be adapted in cases where these conditions are not fully met. We have already illustrated one method of circumventing the last condition; in the revised version of our original transportation problem, it was necessary to introduce a "dummy warehouse" and a suitable set of corresponding transportation costs. This technique made it possible to modify the revised problem so that it would conform to the third condition.

68. The Simplex Method. The "Simplex method" is considered to be a general procedure for linear programming problems. Although the mathematics involved is not difficult, the amount of arithmetical calculation required can be very great. Problems involving only a few variables can be worked out manually or with a desk computer but it is usually necessary to resort to automatic computers for the solution of most practical problems. The procedure used in solving transportation problems is better suited for manual calculation than the standard Simplex method. Consequently, if it is possible to make a linear programming problem conform to the criteria for a transportation problem, it is generally advisable to do so. In this section we illustrate the Simplex method by working out a typical linear programming problem based on a small number of variables. Let us say that x_1, x_2, and x_3 are commodities that can be bought and sold at a profit. However, there are certain constraints placed upon the purchase of these commodities, expressed by the following system of inequalities:

$$x_1 + 2x_2 + 4x_3 \leq 15 \qquad (5\text{-}38)$$

$$4x_1 + 3x_2 + 2x_3 \leq 20 \qquad (5\text{-}39)$$

where $x_1 \geq 0 \quad x_2 \geq 0 \quad x_3 \geq 0$

The profit which is to be maximized is defined by the expression:

$$P = 5x_1 + 4x_2 + 6x_3 \qquad (5\text{-}40)$$

Thus we have a problem which entails the maximization of a linear function subject to a system of linear restraints in the form of inequalities.

In order to solve this problem we must first convert the given inequalities to equalities. This is accomplished by the introduction of dummy variables x_4 and x_5 into expressions (5-38) and (5-39) above, which then gives us:

$$x_1 + 2x_2 + 4x_3 + x_4 = 15 \qquad (5\text{-}41)$$

$$4x_1 + 3x_2 + 2x_3 + x_5 = 20 \qquad (5\text{-}42)$$

where $x_4 \geq 0 \quad x_5 \geq 0$

Since we have only two equations (5-41) and (5-42) we know that the solution to this problem will yield positive values for only two of the five variables. We assume, therefore, arbitrary zero values for the remaining three variables. Now we can set up a *Simplex* tableau from which we can proceed to the solution. First, we select two of the five variables and give them positive values. These are now considered to be *basic variables*. The remaining variables are considered to be of zero value and are termed *non-basic variables*.

We may select our variables arbitrarily in this case, since we do not expect to obtain the optimal solution immediately. For the first tableau to be developed, we will consider x_4 and x_5 to be basic variables and will consider the others to be non-basic. Obviously this selection does not return the greatest profit, for neither of these variables occur in equation (5-40) above. However, this selection does provide a *possible* basis for our first tableau. We now solve equations (5-41) and (5-42) for x_4 and x_5 respecitvely and obtain

$$x_4 = 15 - x_1 - 2x_2 - 4x_3 \qquad (5\text{-}43)$$

$$x_5 = 20 - 4x_1 - 3x_2 - 2x_3 \qquad (5\text{-}44)$$

Next we restate the profit function:

$$P = 0 + 5x_1 + 4x_2 + 6x_3 \qquad (5\text{-}45)$$

From equations (5-43), (5-44) and (5-45) we are now able to construct Tableau I. The framework for this is shown below in Tableau

5-Ia.   The basic variables are written to the left of the tableau and the non-basic variables are written above. The elements in the first column (below C) will be constants.

TABLEAU 5-Ia
Non-basic Variables

	C	x_1	x_2	x_3
P				
Basic x_4				
Variables x_5				

We now fill in the tableau by writing the coefficients of the expressions of the basic variables in terms of the non-basic variables.   For example, the first row in the tableau appears as shown in Tableau 5-Ib.

TABLEAU 5-Ib

	C	x_1	x_2	x_3
P	0	5	4	6
x_4				
x_5				

These numbers were obtained by taking the coefficients from equation (5-45) writing them in the appropriate column.   The second and third rows were obtained from equations (5-43) and (5-44) in the same way. The first tableau is shown below (Tableau 5-Ic) in completed form.

TABLEAU 5-Ic

	C	x_1	x_2	x_3
P	0	5	4	6
x_4	15	-1	-2	-4
x_5	20	-4	-3	-2

If we consider the "C" above the first column to have the value 1, we may say that any basic variable appearing to the left of the tableau is equal to the sum of the products formed by multiplying each element in the designated row by its corresponding column heading.   Thus for x_4 we would obtain $15(1) + (-1)(x_1) + (-2)(x_2) + (-4)(x_3)$ which is equivalent to equation (5-43) above.   Furthermore, since we have designated x_1, x_2, and x_3 as non-basic variables, Tableau 5-Ic tells us that under these circumstances: $P = 0$; $x_4 = 15$; and $x_5 = 20$.

Our object is to increase the value of P and this is accomplished by interchanging a basic variable with a non-basic variable.   Each interchange will result in a completely new tableau and the process of interchanging is continued until all elements in the first row, except for the one directly below C, are preceded by a minus sign.   If one of the elements in the first row and beneath a non-basic variable is positive, it indicates that the profit could be increased by changing this particular non-basic variable to a basic variable.   In order to find the optimal solution it is necessary to conduct a number of "pivoting" operations. The pivot is that element in the tableau which lies at the intersection of the row of the basic variable and the column of the non-basic variable that are to be interchanged.

To determine the pivot we proceed as follows.   First, we decide upon the pivot column by choosing that column which has the *largest positive top element*. In Tableau 5-Ic we would then select column x_3 which is headed by $+6$.   To determine the pivot row, we take each negative element in the chosen column (x_3) and divide it into the corresponding element in the first column.   That element which yields the quotient having the smallest absolute value must be chosen as the pivot.   In Tableau 5-Ic, the negative values in column x_3 are -4 and -2. Dividing these into the corresponding elements in the first column we obtain $(-15/4)$ and (-10) which have the respective absolute values of $(15/4)$ and (10).   The former is obviously the smaller quotient.   Consequently our first pivot is the number -4 lying at the intersection of column x_3 and row x_4.   We now wish to create a new tableau in which x_3 is a basic variable and x_4 is a non-basic variable.   Thus we derive Tableau 5-II by employing the following procedure.

First replace the old pivot element with its reciprocal.   (Tableau 5-IIa)

TABLEAU 5-IIa

	C	x_1	x_2	x_4
P	.	.	.	.
x_3	.	.	.	$-1/4$
x_5	.	.	.	.

Second, the new elements in the same row as the pivot are obtained by dividing the original elements in that row (15; -1; -2) by the old pivot (-4) and changing the sign of the resultant quotients.

Third, the new elements in the same column as the pivot are obtained by dividing the original elements in that column (6; -2) by the old pivot (-4). The results of these three steps are shown below in Tableau 5-IIb.

TABLEAU 5-IIb

	C	x_1	x_2	x_4
P	·	·	·	$-3/2$
x_3	$15/4$	$-1/4$	$-1/2$	$-1/4$
x_5	·	·	·	$1/2$

Finally the remaining elements are found by forming a rectangle using as corners, one of the as yet blank spaces in the new tableau, the pivot, and the two elements in the tableau that are necessary to complete this rectangle. Then, for each remaining new element we have New element = corresponding old element −

$$\frac{\text{(product of complementary corner elements)}}{\text{old pivot}}$$

For example to obtain the new element in the upper left-hand corner of Tableau 5-II we take the original element (0) and subtract from it the product of (6) and (15) divided by the old pivot (-4). (To find these original elements refer back to Tableau 5-Ic). Thus we have

$$0 - \frac{(6)(15)}{(-4)} = \frac{90}{4} = \frac{45}{2}$$

Note that the new element was obtained by employing the elements from Tableau 5-Ic that form the rectangle designated in Tableau 5-IIc below

TABLEAU 5-IIc

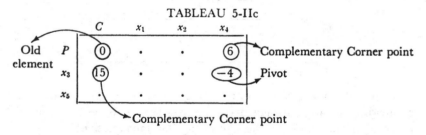

After performing similar operations to determine the remaining new elements, we obtain Tableau 5-IId, which is in completed form.

TABLEAU 5-IId

	C	x_1	x_2	x_4
P	45/2	7/2	1	−3/2
x_3	15/4	−1/4	−1/2	−1/4
x_5	25/2	(−7/2)	−2	1/2

From this tableau we see that, with $x_3 = \frac{15}{4}$ and $x_5 = \frac{25}{2}$, we can obtain a profit of $\frac{45}{2}$. In the first row below the non-basic variables we still have two positive values, $\frac{7}{2}$ and 1. This indicates that we must find a new pivot to improve the solution given in Tableau 5-IId. Selecting the pivot by the same means as before, we find that $-\frac{7}{2}$ circled in Tableau 5-IId, is the correct choice. This indicates that our next step is to make x_1 a basic variable in place of x_5.

Then, by duplicating the first 3 steps described above, we obtain Tableau 5-IIIa from Tableau 5-IId

TABLEAU 5-IIIa

	C	x_1	x_2	x_4
P		−1		
x_3		1/14		
x_5	25/7	−2/7	−4/7	1/7

To complete Tableau 5-III, we find the remaining new elements by employing the appropriate rectangles in Tableau 5-IId. This produces Tableau 5-IIIb which is shown in completed form.

TABLEAU 5-IIIb

	C	x_5	x_2	x_4
P	35	−1	−1	−1
x_3	20/7	1/14	−5/14	−2/7
x_1	25/7	−2/7	−4/7	1/7

At this point we have x_3 and x_1 as basic variables with $x_3 = 20/7$ and $x_1 = 25/7$; our non-basic variables are x_5, x_2, and x_4 with $x_5 = x_2 = x_4 = 0$. Our profit under these circumstances is 35 and since there are no more positive values at the intersections of row P and the three columns of the non-basic variables, we have obtained the optimal solution to this problem. The values $x_3 = 20/7$ and $x_1 = 25/7$ satisfy our original system of inequalities and at the same time yield the maximum profit of 35.

However, it is often necessary that the result of a linear programming problem be expressed in integer form. There is no simple procedure for such cases but usually the nearest integers will produce close approximations to the best solution. In this problem, we could round $x_1 = 25/7$ to $x_1 = 3$ and $x_3 = 20/7$ to $x_3 = 3$ and the profit would drop from 35 to 33 as a result.

69. The Dual Simplex Method. For every linear programming problem there exists an associated complementary problem. That is, for every maximization problem there is a corresponding minimization problem and *vice versa*. The significance of this phenomenon lies in the symmetry of the two problems. Let us regard the problem of maximization of profit as the primal problem. Thus we have:

Primal Problem (Maximization)

$$\text{Max. } P = Ax_1 + Bx_2 + Cx_3 \tag{5-46}$$

subject to the constraints

$$\left.\begin{array}{l} a_{11}x_1 + a_{12}x_2 + a_{13}x_3 \le D \\ a_{21}x_1 + a_{22}x_2 + a_{23}x_3 \le E \\ x_i \ge 0 \ (i = 1, 2, 3) \end{array}\right\} \tag{5-47}$$

Then we have a corresponding problem of minimization (e.g., minimization of cost) which we define as the "dual problem."

Dual Problem (Minimization)

$$\text{Min } Z = Dy_1 + Ey_2 \tag{5-48}$$

subject to the constraints

$$\left.\begin{array}{l} a_{11}y_1 + a_{21}y_2 \ge A \\ a_{12}y_1 + a_{22}y_2 \ge B \\ a_{13}y_1 + a_{23}y_2 \ge C \\ y_i \ge 0 \ (i = 1, 2) \end{array}\right\} \tag{5-49}$$

The tableaus corresponding to these two problems appear as follows where s's and t's are the dummy variables of the primal and dual problem respectively.

P	0	x_1 A	x_2 B	x_3 C
s_1	D	$-a_{11}$	$-a_{12}$	$-a_{13}$
s_2	E	$-a_{21}$	$-a_{22}$	$-a_{23}$

Primal Tableau

Z	0	y_1 $-D$	y_2 $-E$
t_1	$-A$	a_{11}	a_{21}
t_2	$-B$	a_{12}	a_{22}
t_3	$-C$	a_{13}	a_{23}

Dual Tableau

In matrix terminology (see Chapter 2) the dual tableau is the negative transpose of the primal tableau.   That is, to obtain the dual from
the primal, we make each row in the primal a column in the dual and
reverse all the signs.   If we now obtain the optimal solution of the
primal by employing the Simplex method, the solution, of course will
be given by the first column of the tableau.   However, *the first row* in the
solution of the primal problem will also yield the optimal solution to the
dual problem.   This concept of duality plays a very significant role in
the solution of linear programming problems.   For example, in the
preceding section we demonstrated the Simplex method on a maximization problem where all the constraints were linear inequalities of the
type of Expression (5-47)

$$a_{11}x_1 + a_{12}x_2 + \cdots + a_{1n}x_n \leq K_1$$

How do we deal with a minimization problem where the constraints
are of the type?

$$a_{11}x_1 + a_{12}x_2 + \cdots a_{1n}x_n \geq K_1 \tag{5-50}$$

This may, at first, seem to be a deceptively easy problem, but let it
suffice to say that it is not easy.   In fact, to solve a minimization
problem directly by the Simplex method, it is necessary to introduce
twice as many dummy variables as the number needed for solving a
comparable maximization problem.   For this reason it is often expedi-
ent to solve the dual of a minimization problem, which is a maximization problem whose optimal solution will contain the answer to the
original (primal) minimization problem.

Let us now consider an elementary example which will illustrate the
effectiveness of this method.

Example.   Minimize $3x_1 + x_2 = Z$ (5-51)
subject to the constraints

$$\left. \begin{array}{l} 2x_1 + 3x_2 \geq 20 \\ 5x_1 + 3x_2 \geq 24 \end{array} \right\} \tag{5-52}$$

If we first set these expressions up in the form of a minimization problem we obtain

$$2x_1 + 3x_2 - s_1 = 20 \atop 5x_1 + 3x_2 - s_2 = 24 \} \qquad (5\text{-}53)$$

where s_1 and s_2 are dummy variables.

Transposing we obtain

$$s_1 = -20 + 2x_1 + 3x_2 \atop s_2 = -24 + 5x_1 + 3x_2 \} \qquad (5\text{-}54)$$

At this point we may construct an artificial minimization tableau (Tableau 5-IV)

TABLEAU 5-IV

		x_1	x_2
z	0	−3	−1
s_1	−20	2	3
s_2	−24	5	3

This tableau represents the primal problem.   Note that the signs in the first row were reversed.   This is a necessary step in constructing the artificial tableau of a *minimization* problem.   To obtain the dual problem we derive the following expressions from those given in the primal problem.

Maximize $P = 20y_1 + 24y_2,$ $\qquad\qquad\qquad\qquad (5\text{-}55)$

subject to the constraints

$$2y_1 + 5y_2 \leq 3 \atop 3y_1 + 3y_2 \leq 1 \} \qquad (5\text{-}56)$$

Introducing the dummy variables t_1 and t_2 we obtain the following equalities

$$2y_1 + 5y_2 + t_1 = 3 \atop 3y_1 + 3y_2 + t_2 = 1 \} \qquad (5\text{-}57)$$

Transposing, we obtain

$$t_1 = 3 - 2y_1 - 5y_2 \atop t_2 = 1 - 3y_1 - 3y_2 \} \qquad (5\text{-}58)$$

Now we may construct Tableau 5-V which represents the dual problem.

TABLEAU 5-V

		y_1	y_2
P	0	20	24
t_1	3	-2	-5
t_2	1	-3	$\boxed{-3}$

Note that Tableau 5-V is the negative transpose of Tableau 5-IV.   We now proceed to solve the dual problem by the Simplex method.   To find the first pivot element we select column y_2, which has the largest positive top element, and row t_2, since the quotient $\mid 1/3 \mid$ is less than the quotient $\mid 3/5 \mid$.*   The correct pivot element is circled in Tableau V.   Pivoting on the designated element we obtain Tableau 5-VI.

TABLEAU 5-VI

		y_1	t_2
P	8	-4	-8
t_1	4/3	3	5/3
y_2	1/3	-1	$-1/3$

Since there now remain only negative numbers in the first row in the columns headed by non-basic variables (y, and t_2) we have obtained an optimal solution to the dual problem.   The first column in Tableau 5-VI provides the optimal solution to the stated maximization problem.   However, the first row of Tableau 5-VI yields the optimal solution to the primal minimization problem.   Thus we have

$$\left.\begin{array}{l} s_1 = -y_1 = 4 \\[6pt] x_2 = -t_2 = 8 \\[6pt] x_1 = \quad s_2 = 0 \quad \text{Min. } Z = 8 \end{array}\right\} \qquad (5\text{-}59)$$

This result may be verified by trying other combinations in the primal problem.   Keeping this problem in mind and looking back to the symbolic representations of the primal and the dual tableau which are juxtaposed at the beginning of this section, we can now verify that,

*The vinculum denotes absolute value.

when the optimal solution to one of the two problems has been obtained, we obtain the following equalities.

$$\left.\begin{array}{c} x_n = -t_n \\ y_n = -s_n \\ P = Z \end{array}\right\} \qquad (5\text{-}60)$$

This concept of duality has had great significance in the development of linear programming. Besides its importance for the application illustrated in this section, it has also been employed in the solution of two-person zero-sum games and in the development of the techniques of integer programming, two topics which are considered further in the following chapter.

Problems for Solution

1. Solve graphically, locating the corner point of the polygon defined by the inequalities

$$x \leq 4y$$
$$x \geq 2y$$
$$y \geq -7/4\,x + 8$$
$$6y \leq -x + 48$$

which gives the maximum value to the linear expression $3y - x$.

2. Maximize $2y - x$ subject to the constraints

$$y \geq 4$$
$$y \leq -4/7\,x$$
$$7y \leq -2x + 6$$

Solve graphically.

3. Minimize $5x - 4y$ subject to the constraints

$$y \leq 14$$
$$x \leq 13$$
$$y \geq x$$
$$4y \leq 3x + 20$$
$$6y \geq -x + 5y$$
$$3y \leq 4x$$

Solve graphically.

4. Given the following table of shipping costs per ton of material between a set of 3 factories and 5 warehouses (1 dummy warehouse), find the minimum cost of allocating the material. (M is a very large number)

		Warehouses					Factory
		W_1	W_2	W_3	W_4	Z	Excess
	F_1	.08	.06	.10	.15	M	90
Factories	F_2	.12	.09	.07	.08	.00	115
	F_3	.13	.13	.10	.07	.00	55
Warehouse Capacity		60	60	60	60	20	260

Costs are in cents per ton. Excess and capacity are measured in tons.

5. The margin table below records the profits that are possible when one ton of material is sold by F_i to D_j where F_i are factories and D_j are dealers. The supply of material on hand at the factories is 135 tons which exactly coincides with the demand of the dealers. Hence there is no need to introduce a dummy variable. Find the set of allocations which produces the *maximum* profit for the factories.

		Dealers				Factory
		D_1	D_2	D_3	D_4	Supply
	F_1	.25	.30	.20	.20	100
Factories	F_2	.30	.25	.15	.10	20
	F_3	.10	.35	.05	.30	15
Dealer Demand		30	30	30	45	135

Supply and demand are measured in tons and margin is measured in cents per ton.

6. Solve by the Simplex method.

Maximize $5x_1 + x_2 + x_3 = P$

subject to the constraints

$$x_1 + 5x_2 + 4x_3 \leq 36$$
$$7x_1 + x_2 + 2x_3 \leq 40$$

let the dummy variables be basic in the first tableau.

7. Solve by the Simplex method

Maximize $3x + 4y + 2z = P$

subject to the constraints

$$x + y + z \leq 20$$
$$2x + z \leq 12$$

Chapter 6

COMBINATORIAL MATHEMATICS

Combinatorial mathematics is the name applied to an extremely great variety of methods and techniques. Thus, combinatorial topology is that branch of topology which studies geometric forms by decomposing them into the simplest geometric figures which adjoin each other in a regular fashion. However, combinatorial mathematics is not restricted to topology, but has been extended to many algebraic topics, including that of linear programming discussed in the preceding chapter. For while combinatorial methods have long been known to mathematicians, they have in recent years been found useful in solving a wide variety of practical problems.

In the foregoing chapter we introduced some of the basic concepts and applications of linear programming. We stressed the transportation problem procedure because of its mathematical simplicity and because of its adaptibility to manual calculations.

Finding optimum combinations is only one aspect of combinatorial mathematics, but it is this aspect that we have chosen to introduce in the preceding chapter and which is more fully investigated in the first section of the present chapter.

70. Integer Programming. The Simplex method, as introduced in the preceding chapter, provides an efficient method for solving most linear programming problems. However, one of the most severe limitations of the original Simplex method is that the resulting optimal solutions are always exact, and therefore often fractional. At first thought it hardly seems proper to refer to this characteristic as a limitation, but there are a great many linear programming problems which demand integer solutions. As we saw in the preceding chapter, the Simplex method, as often as not, produces a set of optimal solutions which may include unwieldy fractions. Such a set of solutions is optimal from the mathematical standpoint but is still unsatisfactory when the problem prescribes a solution in the form of integers. In this section we will deal with the subject of integer programming, which does supply a technique for obtaining the optimal integer solution for linear programming problems.   It should be stressed here that the study of integer programming is a very important branch of linear programming.

The methods that are currently available for dealing with this kind of problem are generally too involved for manual calculation and are not infallible.   Consequently this subject area is one of the major concerns of modern mathematicians, who are seeking to revise and improve upon the existing methods.   The technique presented in this chapter demonstrates the general method of attacking an integer programming problem.   It is subject to the limitation that if the problem is very complex, feasible solutions are not available.

The optimal integer solution for a linear programming problem is illustrated by the following example.   Consider the problem stated in the customary form:

$$\text{Maximize } P = 6x_1 + 4x_2 + x_3 \qquad (6\text{-}1)$$

subject to the constraints

$$\left.\begin{array}{l} 3x_1 + 2x_2 \leq 10 \\ x_1 + 4x_2 \leq 11 \\ 3x_1 + 3x_2 + x_3 \leq 13 \end{array}\right\} \qquad (6\text{-}2)$$

Before we can obtain the optimal integer solution, it is necessary to find the exact optimal solution by employing the Simplex method.   Thus, by introducing dummy variables we obtain

$$\left.\begin{array}{l} 3x_1 + 2x_2 + t_1 = 10 \\ x_1 + 4x_2 + t_2 = 11 \\ 3x_1 + 3x_2 + x_3 + t_3 = 13 \end{array}\right\} \qquad (6\text{-}3)$$

Transposing these equations and solving for the dummy variables, we obtain

$$\left.\begin{array}{l} t_1 = 10 - 3x_1 - 2x_2 \\ t_2 = 11 - x_1 - 4x_2 \\ t_3 = 13 - 3x_1 - 3x_2 - x_3 \end{array}\right\} \qquad (6\text{-}4)$$

We are now able to construct an initial tableau based on t_1, t_2, and t_3 as basic variables. This gives us Tableau 6-I.

TABLEAU 6-I

		x_1	x_2	x_3
P	0	6	4	1
t_1	10	⟨-3⟩	-2	0
t_2	11	-1	-4	0
t_3	13	-3	-3	-1

For our first pivot we select -3 (circled above) which lies on the intersection of row t_1 and column x_1. The pivot column was chosen by selecting the column having the largest positive top element. Note that the first column is never eligible for the pivot column since its elements express the constant values in a series of equations. Thus having selected column x_1, headed by the top element 6, it is then necessary to determine the pivot row. To accomplish this each element in column x_1 preceded by a minus sign is divided into its corresponding element in the first column. The resultant quotient having the lowest absolute value designates the pivot row. In this case we obtained quotients with absolute values $\frac{10}{3}$, 11, and $\frac{13}{3}$. The fraction $\frac{10}{3}$ is smallest in absolute value and was derived from row t_1; thus t_1 is the pivot row. Then we construct Tableau 6-II by following the rules described in the preceding chapter but which are repeated here for convenience. First, the new pivot is the reciprocal of the old pivot, i.e. $-\frac{1}{3}$. Second, the remaining elements in the pivot row (t_1 of Tableau 6-I) are found by dividing the original elements in that row by the old pivot (-3) and reversing the sign of the resulting products. Third, the remaining elements in the pivot column (x_1 of Tableau 6-I) are found by dividing the original elements in that column by the old pivot (-3). Finally, the remaining elements in the new tableau are found by subtracting from the original element the quotient determined by dividing the product of the complementary corner elements in the original tableau by the old pivot.

TABLEAU 6-II

		t_1	x_2	x_3
P	20	-2	0	1
x_1	10/3	-1/3	-2/3	0
t_2	23/3	1/3	-10/3	0
t_3	3	1	-1	⟨-1⟩

Our solution, at this point, is still not optimal, so another tableau must be constructed. We find that (-1) at the intersection of t_3 and x_3 should be the new pivot. Using this element as our new pivot we obtain Tableau 6-III.

TABLEAU 6-III

	t_1	x_2	t_3
P	23	-1	-1	-1
x_1	3-1/3	$-1/3$	$-2/3$	0
t_2	7-2/3	1/3	-3-1/3	-1
x_3	3	1	-1	-1

This tableau does provide the exact optimal solution to the problem, i.e. $P = 23$; $x_1 = 3\frac{1}{3}$; $t_2 = 7\frac{2}{3}$; $x_3 = 3$; $t_1 = t_3 = x_2 = 0$. However, this is not an integer solution, for two of the elements in the first column ($3\frac{1}{3}$ and $7\frac{2}{3}$) contain fractions. To obtain the optimal integer solution to this problem, it is necessary to introduce an additional constraint, which is referred to as the Gomory constraint after Ralph E. Gomory who developed this method for obtaining integer solutions. The new constraint, when determined, will add an additional row at the bottom of Tableau 6-III.

The Gomory constraint is based on a row already existing in the optimal solution and this row is the one whose first element has the greatest fractional part. Thus, upon examining Tableau 6-III, we find that the third row, i.e. row t_2, has the first element with the greatest fractional part, i.e. $\frac{2}{3}$. The first element in the Gomory constraint is taken as the negative of this fraction, i.e. $\frac{2}{3}$. The first element in the Gomory constraint is the negative of this fraction, i.e. $-\frac{2}{3}$. To find the remaining elements in the Gomory constraint we first take the fractional parts of the remaining elements in the selected row (t_2). These are as follows:

$$1/3 \qquad -1/3 \qquad 0 \qquad\qquad (6\text{-}5)$$

Then we write the negative of each of these elements

$$-1/3 \qquad 1/3 \qquad 0 \qquad\qquad (6\text{-}6)$$

Then we write the above line again with the positive and 0 elements unchanged, but with new values calculated for the negative elements by substituting them in the expression $-1 + (1 + a)$ where a is the negative element. Thus we obtain from the above line

$$-1 + \left(1 + \left(-\frac{1}{3}\right)\right) = -1 + \frac{2}{3} \qquad\qquad (6\text{-}7)$$

where the second and third elements ($\frac{1}{3}$ and 0) are unchanged, but the first element $-1 + \frac{2}{3}$ is obtained by substituting the $-\frac{1}{3}$ from the previous line in the expression $-1(1 + a)$ giving $-1 + (1 + -\frac{1}{3}) = -1 + \frac{2}{3}$. Finally we drop all -1's from this set of numbers. Consequently, we have

$$2/3 \qquad 1/3 \qquad 0 \qquad\qquad (6\text{-}8)$$

which are the coefficients to be used for the remaining elements of the Gomory constraint. Expressed symbolically, where

$$a_{i0} + a_{i1} + a_{i2} + \cdots + a_{in} \qquad\qquad (6\text{-}9)$$

is the row selected as the basis for the Gomory constraint; we define the Gomory constraint itself as

$$-f_{i0} + \sum_{j=1}^{n} f_{ij} \qquad\qquad (6\text{-}10)$$

where f_{i0} is the fractional part of the element a_{i0} and where $f_{ij}(j = 1, \cdots, n)$ are the positive fractional parts of $-a_{ij}(j = 1, \cdots, n)$. For the problem under discussion $a_{i0} = 7\frac{2}{3}$ and $-f_{i0} = -\frac{2}{3}$. Likewise $a_{i1} = \frac{1}{3}$ and $f_{i1} = \frac{2}{3}$; $a_{i2} = -3\frac{1}{3}$ and $f_{i2} = \frac{1}{3}$; $a_{i3} = 0$ and $f_{i3} = 0$. Thus we may now construct Tableau 6-IV which is simply Tableau 6-III with the addition of g_1, the Gomory constraint.

TABLEAU 6-IV

		t_1	x_2	t_3
P	23	-1	-1	-1
x_1	3-1/3	$-1/3$	$-2/3$	0
t_2	7-2/3	1/3	-3-1/3	0
x_3	3	1	-1	-1
g_1	$-2/3$	2/3	1/3	0

Since Tableau 6-III already provided us with an optimal solution, so that the coefficients of the non-basic variables in the first row are all negative, we are not able to select a pivot by the usual method. However, not all of the elements in the first column are positive. From this observation we know that the solution portrayed in Tableau 6-IV (unlike that of Tableau 6-III) is not optimal. Since we are not able to optimize the primal problem by the usual method, we now attempt to optimize the dual problem. To obtain the tableau for the dual problem we construct the negative transpose of Tableau 6-IV. This is shown below as Tableau 6-V.

TABLEAU 6-V

	(P')	x'_1	t'_2	x'_3	g'_1
z	-23	$-3\text{-}1/3$	$-7\text{-}2/3$	-3	$2/3$
t'_1	1	$1/3$	$-1/3$	-1	$\boxed{-2/3}$
x'_2	1	$2/3$	$3\text{-}1/3$	1	$-1/3$
t'_3	1	0	0	1	0

Now we may proceed by following the standard Simplex method·
First we determine the pivot element which is $(-\frac{2}{3})$ at the intersection
of row t'_1 and column g'_1. Using this element as a pivot we obtain
Tableau 6-VI.

TABLEAU 6-VI

	(P')	x'_1	t'_2	x'_3	t'_1
z	-22	-3	-8	-4	-1
g'_1	$3/2$	$1/2$	$-1/2$	$-3/2$	$-3/2$
x'_2	$1/2$	$1/2$	$7/2$	$3/2$	$1/2$
t'_3	1	0	0	1	0

It is now evident that we have again reached an optimal solution, for
all of the coefficients of the non-basic variables in the first row are
negative. Consequently, we are unable to determine another pivot
element. However, all of the elements in the first row are integers and
we know that the first row in the optimal solution of the dual problem
provides the optimal solution of the primal problem. Thus we have
obtained the optimal integer solution. To return to the primal
tableau, we construct the negative transpose of Tableau 6-VI. This
construction is based on the obvious theorem that "the dual of the dual
is the primal." The optimal integer solution to our original problem
is illustrated below by Tableau 6-VII.

TABLEAU 6-VII

		g_1	x_2	t_3
P	22	$-3/2$	$-1/2$	-1
x_1	3	$-1/2$	$-1/2$	0
t_2	8	$1/2$	$-7/2$	0
x_3	4	$3/2$	$-3/2$	-1
t_1	1	$3/2$	$-1/2$	0

Thus we have:

$$x_1 = 3 \qquad t_1 = 1$$
$$x_3 = 4 \qquad t_2 = 8$$
$$g_1 = x_2 = t_3 = 0$$

Max $P_I = 22$ where P_I is P based on an integer program.

It is not always possible to obtain an integer solution by the addition of a single Gomory constraint. If the solution resulting from the insertion of a Gomory constraint contains fractional parts, we must continue to devise additional constraints by repeating the method illustrated above. Usually this technique will produce an integer solution after a finite number of steps. However, if the problem is degenerate in some aspect, it may be difficult or impossible to solve it without employing more elaborate techniques.

71. Linear Programming and Game Theory. Linear programming and game theory are both comparatively recent developments in modern mathematics. However, both of these disciplines are now regarded as very important tools for the solution of practical problems in business and industry. At first thought there does not seem to be any relation between them, other than their use on occasion to solve the same type of problems. That is, our first impression might be that linear programming and game theory are associated only because both are modern mathematical techniques having practical applications based on the optimization of one of a set of variables.

In fact, however, there is a very important mathematical relationship here. The primary purpose of this section to illustrate this relationship, showing how a two-person zero-sum game may be solved by the techniques of linear programming.

In the chapter on the theory of games (Chapter 4) we employed the following game as an example of a two-person zero-sum game in which each player has at his disposal three strategies.

		Player B		
		1	2	3
	1	4	−1	0
Player A	2	0	1	3
	3	−2	4	1

We will now solve this same problem by treating it as an exercise in linear programming.

Our first objective will be to determine B's optimum grand strategy. The reason for this choice will become apparent as we get deeper into the problem. We know that we are looking for a grand strategy that will satisfy the criterion

$$y_1 + y_2 + y_3 = 1 \qquad (6\text{-}11)$$

(y_i = the element of odds corresponding to B's ith strategy)

Furthermore, we know that, if the grand strategy is optimal, the following set of constraints must also be satisfied.

$$\left.\begin{array}{r}4y_1 - y_2 \quad\quad \leq v \\ y_2 + 3y_3 \leq v \\ -2y_1 + 4y_2 + y_3 \leq v\end{array}\right\} \quad\quad (6\text{-}12)$$

where v is the value of the game and since B is the minimizing player, we wish to satisfy the above constraints for the smallest possible v.

In order to solve this game by the techniques of linear programming we must first ensure that the value of the game is positive. We know that if we add the same constant to each element in the game, we will also increase the value of the game by that amount without altering the grand strategies of the two players. Thus we now add $(+3)$ to each element in the game matrix.

		Player B		
		1	2	3
	1	7	2	3
Player A	2	3	4	6
	3	1	7	4

The value of this game must now be positive, for there are no negative elements in the game matrix. To change the original set of constraints into a set of equations, we employ the coefficients in the revised game matrix and introduce the new variables $y_i/v = \Upsilon_i$. We also denote $1/v$ as V and introduce the customary dummy variables, obtaining:

$$\left.\begin{array}{r}7\Upsilon_1 + 2\Upsilon_2 + 3\Upsilon_3 + t_1 = 1 \\ 3\Upsilon_1 + 4\Upsilon_2 + 6\Upsilon_3 + t_2 = 1 \\ \Upsilon_1 + 7\Upsilon_2 + 4\Upsilon_3 + t_3 = 1 \\ \Upsilon_i \geq 0 \; (i = 1, 2, 3) \quad\quad t_i \geq 0 \; (i = 1, 2, 3)\end{array}\right\} \quad\quad (6\text{-}13)$$

To obtain this set of constraints, we have, in effect divided both sides of the old set of constraints by v and introduced the dummy variables t_i $(i = 1, 2, 3)$. The coefficients, as stated above, were taken from the revised game matrix. Instead of minimizing v, we will now

$$maximize \frac{y_1 + y_2 + y_3}{v} = V(y_1 + y_2 + y_3) = \Upsilon_1 + \Upsilon_2 + \Upsilon_3 = V'.$$

$$\left(where \; V' = \frac{1}{v + 3}\right)$$

Transposing the above equations we obtain

$$\left.\begin{array}{l} t_1 = 1 - 7\Upsilon_1 - 2\Upsilon_2 - 3\Upsilon_3 \\[4pt] t_2 = 1 - 3\Upsilon_1 - 4\Upsilon_2 - 6\Upsilon_3 \\[4pt] t_3 = 1 - \Upsilon_1 - 7\Upsilon_2 - 4\Upsilon_3 \\[4pt] V' = 0 + \Upsilon_1 + \Upsilon_2 + \Upsilon_3 \end{array}\right\} \qquad (6\text{-}14)$$

The original game can now be represented by the following Simplex tableau.

TABLEAU 6-VIII

		Υ_1	Υ_2	Υ_3
V'	0	1	1	1
t_1	1	$\boxed{-7}$	-2	-3
t_2	1	-3	-5	-6
t_3	1	-1	-7	-4

Since all the coefficients of the non-basic variables in the first row are $+1$, we may select the first pivot column arbitrarily. If we select Υ_1 as the pivot column we find that t_1 is the correct pivot row and that -7 is the corresponding pivot element, for, when the negative elements in Υ_1 are divided into their corresponding elements in the first column, the quotient $\frac{1}{7}$ has the smallest absolute value. Using this pivot we obtain Tableau 6-IX by the Simplex method.

TABLEAU 6-IX

		t_1	y_2	y_3
V'	1/7	$-1/7$	5/7	4/7
y_1	1/7	$-1/7$	$-2/7$	$-3/7$
t_2	4/7	3/7	$-22/7$	$-33/7$
t_3	6/7	1/7	$\boxed{-47/7}$	$-25/7$

This solution is still not optimal, for two positive values are present in the first row below the signed columns. Thus we select $-\frac{47}{7}$ as the new pivot. This pivot column was selected because it is headed by the highest positive top element. The pivot row was then selected by dividing the negative elements in the pivot column into the corresponding elements in the first column and selecting that element in the pivot row for which the resulting quotient was smallest in absolute value.

Using this pivot we obtain Tableau 6-X by the Simplex method. First, we replace $-\frac{47}{7}$ with its reciprocal. Second, we divide the remaining elements in the pivot row by the pivot element, i.e., $-\frac{47}{7}$ and change the signs of the resultant quotients. Third, we divide the remaining elements in the pivot column by the pivot element. To obtain the remaining elements in Tableau 6-X we employ the formula

new element =

$$\text{old element} - \frac{[\text{product of complementary corner points}]}{\text{pivot}} \qquad (6\text{-}15)$$

For example, we obtain $\frac{11}{47}$ in the upper left-hand corner of Tableau 6-X as follows

$$\frac{1}{7} - \frac{(5/7)(6/7)}{(-47/7)} = \frac{11}{47}$$

TABLEAU 6-X

		t_1	t_3	y_3
V'	11/47	−6/47	−5/47	9/47
y_1	5/47	−7/47	2/47	−13/47
t_2	8/47	17/47	22/47	−143/47
y_2	6/47	1/47	−7/47	−25/47

This solution is not optimal and we continue by selecting a third pivot element, $-\frac{143}{47}$, at the intersection of row t_2 and column y_3. Pivoting on this element we obtain Tableau 6-XI.

TABLEAU 6-XI

		t_1	t_3	t_2
V'	35/143	−14/143	−11/143	−9/143
y_1	13/143	−26/143	0	13/143
y_3	8/143	17/143	22/143	−47/143
v_2	14/143	−6/143	−33/143	25/143

This solution is optimal, for all the elements in the first row heading signed columns are preceded by a minus sign. Thus we are now able to determine the value of the original game and the optimal grand strategies for each player.

The value of the game is equal to the reciprocal of the element in the upper left hand corner of the tableau minus three, the quantity added previously to all the elements in the game matrix. Thus we have

$$
\left.
\begin{aligned}
v &= \frac{1}{V'} - 3 \\[2mm]
v &= \frac{1}{35/143} - 3 \\[2mm]
v &= \frac{143}{35} - \frac{105}{35} = \frac{38}{35}
\end{aligned}
\right\}
\qquad (6\text{-}16)
$$

This figure agrees with the value calculated by standard game procedure on Page 153. Since Tableau 6-XI represents the solution of the game in terms of Player B, we will now derive his optimal grand strategy. We have the general conversion formula

$$
y_i = \Upsilon_i / V' \qquad (i = 1, 2, 3)
$$

Thus we obtain

$$
\left.
\begin{aligned}
y_1 &= \left(\frac{13}{143}\right)\left(\frac{143}{35}\right) = \frac{13}{35} \\[2mm]
y_2 &= \left(\frac{14}{143}\right)\left(\frac{143}{35}\right) = \frac{14}{35} \\[2mm]
y_3 &= \left(\frac{8}{143}\right)\left(\frac{143}{35}\right) = \frac{8}{35}
\end{aligned}
\right\}
\qquad (6\text{-}17)
$$

Player B's optimal grand strategy is $\frac{13}{35} : \frac{14}{35} : \frac{8}{35}$, or to obtain the solution in the more familiar form we multiply each of these figures by 35, which yields a 13 : 14 : 8 strategy (as found by game theory on Page 153). These two expressions for B's optimal grand strategy are equivalent. Finally, to obtain Player A's optimal grand strategy, it is only necessary to refer to the dual of Tableau 6-XI. We are concerned with the elements in the first row of the tableau. If Player A's grand strategy consists of three figures x_i such that

$$
x_1 + x_2 + x_3 = 1
$$

and we have $x_i = t_i / V'$ $\qquad (i = 1, 2, 3)$

This yields:

$$x_1 = \left(\frac{15}{143}\right)\left(\frac{143}{35}\right) = \frac{15}{35}$$

$$x_2 = \left(\frac{9}{143}\right)\left(\frac{143}{35}\right) = \frac{9}{35} \qquad (6\text{-}18)$$

$$x_3 = \left(\frac{11}{143}\right)\left(\frac{143}{35}\right) = \frac{11}{35}$$

Thus A's optimal grand strategy is $\frac{15}{35} : \frac{9}{35} : \frac{11}{35}$ or $15 : 9 : 11$ (as found on Page 153). From this example we have seen that a matrix game may be expressed as and solved as a linear programming problem by employing the dual Simplex method. Thus, games that are too complex to be tractable by the usual methods of game theory can be transformed into a linear programming problem and as such can be solved, if necessary, by an electronic computer.

72. Network Flow Problems. In this section we explore the methods for determining the maximum flow in a given *network*. To define the word network it is first necessary to introduce some new terms. Points specifically designated on a given plane are referred to as *nodes* and a line connecting a pair of nodes is called an *arc*. A collection of nodes and arcs is defined as a *graph* and a series of connected arcs linking a group of nodes is called a *chain*. Two of the nodes of a graph are distinguished from the rest by the designations *source* and *sink*. When each arc is assigned a capacity, we have a *network* and the problem is then to send the maximum flow from source to sink.

Consider Figure 6-1 below in which we have a typical network with specific nodes designated as the source and the sink and a collection of arcs with designated capacities. In this case the capacity at any node can be thought of as infinite, or at least as great enough to accept any amount of liquid that may arrive through its arcs. Liquid is conserved

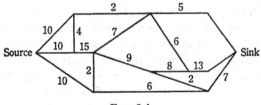

Fig. 6-1.

at all points in the network except at the source and the sink. The source may also be considered to be infinite and the sink may be thought of as having an infinite capacity. Our object is to find the maximum flow that is possible between the source and the sink by assignment of flows to the arcs, the only limitations being, of course, their individual capacities.

Stated symbolically, the conservation of liquid at all points except for the source and the sink can be represented as

$$\sum_i x_{ip} - \sum_j x_{pj} = 0, \qquad p \neq 0, n \qquad (6\text{-}19)$$

where x_{ij} is the flow from point i to point j and p_i and p_j denote points in the network; p_0 and p_n indicate the source and sink respectively. We also have a system of inequalities of the form

$$x_{ij} \leq c_{ij} \qquad (6\text{-}20)$$

which state that flow does not exceed capacity where c_{ij} is the capacity of the arc connecting point i and point j. Flow is represented by the equation

$$f = \sum_i x_{in} = \sum_j x_{0j} \qquad (6\text{-}21)$$

and this is the expression that we wish to maximize. Thus we again have a linear expression that is to be maximized subject to a system of linear constraints.

If the amount of liquid flowing through a given arc is equal to the capacity of that arc, we say that the arc is *saturated*. If an arc is not saturated, the difference between its capacity and the load that it is carrying is called its *residual flow*. A *cut* is a collection of arcs whose removal from the system splits the network into two separate parts, one containing the source, the other containing the sink. If a cut is removed from a network, it follows that no liquid may flow from the source to the sink.

Using this terminology we can say that the maximal flow possible in a given network is less than or equal to the value of any cut, for it is evident that all the liquid flowing from the source to the sink must pass through every cut. Consequently, it must pass through the cut having the smallest capacity. Hence the total flow from source to sink can not exceed the capacity of the smallest cut. In fact, as we shall see, the maximum flow is exactly equal to the capacity of the minimum cut. This phenomenon is known as the max flow min cut theorem and is expressed symbolically as

$$\max f = \min C. \tag{6-22}$$

where C is a cut.

In Figure 6-2 six of the possible cuts in our original network are designated by the lines superimposed on the system. Each set of arcs crossed by one of these lines is a cut and Table 6-1 below gives the capacities of this sampling of cuts.

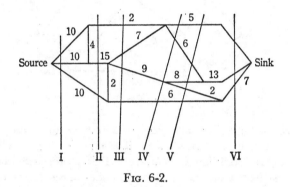

FIG. 6-2.

TABLE 6-1

Cut	I	II	III	IV	V	VI
Capacity	30	27	24	26	27	25

In this table we see that cut III has the minimum capacity, which is 24. Since this cut is the minimum cut in the given network, the maximum flow must also be equal to 24.

To construct the maximum flow, that is, to determine how much flow must be carried in each arc to produce this maximum flow in the network, we begin by setting up an arbitrary network flow. Then if there is a chain from the source to the sink such that none of the included arcs is saturated, we can increase the network flow by adding to each arc in the chain the smallest residual flow of its arcs. However, if there is no chain from source to sink without a saturated arc, we must test the network to ascertain if this is the optimal flow arrangement. In Figure 6-3 we have constructed an arbitrary flow through

the given network.   In this and the subsequent figures the source will
be symbolized by "α" (Alpha) and the sink by "ω" (Omega).   The
remaining nodes are designated by small roman letters.   There are
three distinct chains from source to sink in Figure 6-3.   We have the
routes $\alpha - a - b - \omega$ carrying two units, $\alpha - c - d - e - f - \omega$
carrying eight units, and $\alpha - g - h - \omega$ carrying six units.   This
flow arrangement gives us a total flow of 16 units through the network.
We will now attempt to improve this flow by the "labelling process."

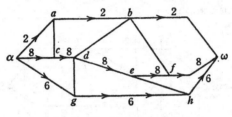

Fig. 6-3.

We begin by labelling all nodes which have an unsaturated arc
connecting directly to the source.   These nodes are labelled (x, α)
where x is the residual flow in that arc.   This expression indicates that
an additional flow x can be drawn from the source to the designated
node.   Thus, in Figure 6-4 above, nodes a, b, and c have been labelled
in this fashion.   The steps in this labelling are as follows.   The arc

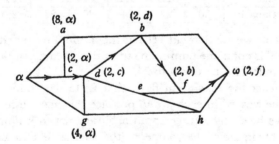

Fig. 6-4.

connecting α and a has a capacity of 10.   After constructing the flow
diagram in Figure 6-3, the arc $\alpha - a$ has a residual flow of 8, giving us
the label $(8, \alpha)$ for node a.

Similarly the arc $\alpha - g$ has a residual flow of its capacity shown in Figure 6-2 minus the flow assumed in Figure 6-3, or $10 - 6 = 4$, so node g is labelled $(4, \alpha)$ in Figure 6-4. Also the arc $\alpha - c$ has a residual flow of $10 - 8 = 2$, so that node c is labelled $(2, \alpha)$.

The next step in labelling depends upon the route from source to sink which we choose to investigate for its availability for additional flow. Since arc $\alpha - c$ is within 2 units of saturation, we choose it as the first arc in a possible route, and decide to try $d - b - f - \omega$ as the remainder of the route. Then we label in order the nodes along this route with this additional 2 units, provided that each arc so connected can accept this available flow. (The advantage of labelling is that if we encounter an arc that cannot take additional flow, i.e., is already saturated, we can try an alternate pathway from the previous labelled node, without starting all the way back at the source.)

We find that arc $c - d$ has a capacity of 15 (Figure 6-2), and is carrying 8 (Figure 6-3); therefore its residual flow is 7, so it can carry 2 more. Accordingly, we label node d, $(2, c)$ in Figure 6-4. The next arc in our trial route is $d - b$. From Figure 6-2, its capacity is 7; from Figure 6-3, it is carrying 0; so its residual flow is 7 and it can carry two more; whence we label node b in Figure 6-4, $(2, d)$. The next arc in the route is $b - f$. Its capacity (Figure 6-2) is 5; it is carrying (Figure 6-3) 3; therefore, it can carry 2 more; so we label node f in Figure 6-4, $(2, b)$. Finally, the arc $f - \omega$ has a capacity (Figure 6-2) of 13; it is carrying (Figure 6-3) 8, so it can carry 2 more, and we label the sink $(2, f)$.

We have thus found that when the arc $\alpha - c$ is saturated by adding 2 units to its flow, the added 2 units can be carried through to the sink by the route $\alpha - c - d - b - f - \omega$. Therefore we prepare a new graph (Figure 6-5) by adding to the flows in the arcs of Figure 6-3 those given by the labelled nodes of Figure 6-4. We have now

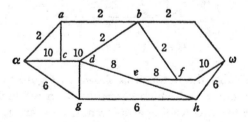

Fig. 6-5.

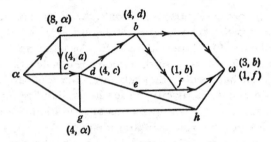

FIG. 6-6.

increased the total flow to 18 units. Now we relabel the network as shown in Figure 6-6. This time we will attempt to minimize the residual flow in the arc $\alpha - a$. At present the residual flow in this arc is 8 and there are only two arcs leading away from node a. Arc $a - b$ is already saturated; thus only arc $a - c$ remains to carry the additional flow and the capacity of this arc is 4. Consequently, we find that it will be impossible to use arc $\alpha - a$ to capacity but it may be possible by starting with $\alpha - a$ to send an additional flow of four units through the system. Through the labelling process we find that we can dispose of these four units by sending three units along the chain $\alpha - a - c - d - b - \omega$ and one unit along the chain $\alpha - a - c - d - b - f - \omega$. After this operation we obtain a new set of flows as illustrated in Figure 6-7 and the total flow has now been increased to 22 units.

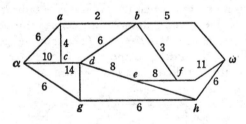

FIG. 6-7.

We again relabel the network as shown in Figure 6-8. This time we investigate arc $\alpha - g$ which currently has a residual flow of four units. Of the two arcs leading away from node g, arc $g - h$ is already saturated and arc $g - d$ has a capacity of 2 units. Thus we know

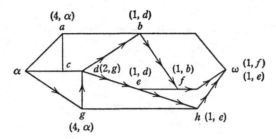

FIG. 6-8.

that we can not reduce the residual flow in arc $\alpha - g$ below 2 but it may be possible to reduce it this far. By again employing the labelling process we find that one unit of flow can be sent along the chain $\alpha - g - d - b - f - \omega$ and one unit of flow can be sent along the chain $\alpha - g - d - e - h - \omega$. We now have a flow of 24 units from the source to the sink of the given network and the flows in each arc are shown in Figure 6-9.

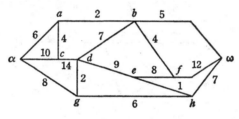

FIG. 6-9.

In Figure 6-10 the arcs that are now saturated are designated by the symbol —|—|—. It appears that we have now obtained the optimal solution, but to check, we attempt to isolate a cut from the given network made up entirely of saturated arcs. The cut that severs the arcs $a - b$, $d - b$, $d - e$, and $g - h$ fits this description. The capacity of this cut is 24 and, in fact, this is the same cut as the one designated "cut III" in Table 6-1 of this section. Thus, we have min cut = max flow = 24 units for this network.

Some networks are too complex to be solved easily by the graphical method. In such cases it is usually more feasible to employ a series of tables in which rows correspond to nodes from which a flow can emerge and columns correspond to nodes that can accept an incoming flow.

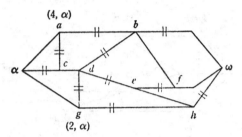

FIG. 6-10.

In the first table, setting the total flow equal to zero, the entries are simply the arc capacities. If two nodes are not connected by an arc the capacity is considered to be zero. Consider the network illustrated in Figure 6-11 and Table 6-2 which corresponds to the arc capacities when the flow is equal to zero.

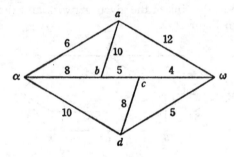

FIG. 6-11.

TABLE 6-2

	α	a	b	c	d	ω	
α	—	6	8	0	10	0	
a	6	—	10	0	0	12	$(6, \alpha)$
b	8	10	—	5	0	0	
c	0	0	5	—	8	4	
d	10	0	0	8	—	5	
ω	0	12	0	4	5	—	$(6, a)$

When a flow is established from the ith node to the jth node, we *subtract* the amount of this flow from the element in the ith row and the jth column. Then we *add* this same amount to the element in the jth row and ith column. Thus we always have the residual flow in an ij cell and the flow from i to j at any time is equal to the original entry in that cell minus the current entry or is equal to the negative difference of the original entry in the *ji* cell and the current entry in that cell.   The labelling process follows the same pattern as before.   In Table 6-2 we have constructed a chain from the source to the sink via node *a* carrying a flow of 6 units.   The results of this initial flow are shown in Table 6-3, along with a new set of labels for the next proposed flow.

TABLE 6-3

	α	a	b	c	d	ω	
α	—	0	8	0	10	0	
a	12	—	10	0	0	6	$(8, b)$
b	8	10	—	5	0	0	$(8, \alpha)$
c	0	0	0	—	8	4	
d	10	0	0	8	—	5	
ω	0	18	0	5	4	—	$(6, a)$

The labels in Table 6-3 suggest a flow of 6 units along the chain $\alpha - b - a - \omega$.   The results of this flow are shown in Table 6-4 with the next proposed flow and so on through Table 6-7.

TABLE 6-4

	α	a	b	c	d	ω	
α	—	0	2	0	10	0	
a	12	—	16	0	0	0	
b	14	4	—	5	0	0	$(2, \alpha)$
c	0	0	0	—	8	4	$(2, b)$
d	10	0	0	8	—	5	
ω	0	24	0	4	5	—	$(2, c)$

TABLE 6-5

	α	a	b	c	d	ω	
α	—	0	0	0	10	0	
a	12	—	16	0	0	0	
b	16	4	—	3	0	0	
c	0	0	7	—	8	2	
d	10	0	0	8	—	5	$(10, \alpha)$
ω	0	24	0	6	5	—	$(5, d)$

TABLE 6-6

	α	a	b	c	d	ω	
α	—	0	0	0	5	0	
a	12	—	16	0	0	0	
b	16	4	—	3	0	0	
c	0	0	7	—	8	2	$(5, d)$
d	15	0	0	8	—	0	$(5, \alpha)$
ω	0	24	0	6	10	—	$(2, c)$

TABLE 6-7

	α	a	b	c	d	ω
α	—	0	0	0	3	0
a	12	—	16	0	0	0
b	16	4	—	3	0	0
c	0	0	7	—	10	0
d	17	0	0	6	—	0
ω	0	24	0	8	10	—

From the last column in Table 6-7 we see that there is no residual flow in any of the arcs joining a node to the sink. Thus we have obtained an optimal solution. The resulting flows and their directions

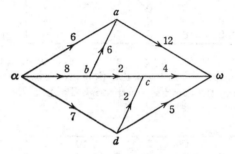

Fig. 6-12.

are illustrated below in Figure 6-12. This figure shows that arcs $a - \omega$, $c - \omega$, and $d - \omega$ are all saturated. This set of arcs is a cut whose capacity is 21. Thus we again have:

$$\text{min cut} = 21 = \text{max flow.}$$

The network flow method may be applied to a number of linear programming problems, including transportation problems which restrict the use of certain routes, scheduling problems, and personnel assignment problems.

73. Trees and Loops. Two special types of graphs, the tree and the loop, have properties that are of great significance in the solution of practical problems. A graph is called a *tree* if it is a network containing no loops, a *loop* being a chain of arcs which originates and terminates at the same node. In Figures 6-13a and 6-13b we have two identical distributions of nodes, but in Figure 6-13a the arcs connecting these nodes are in the form of a tree and in Figure 6-13b these arcs describe a loop.

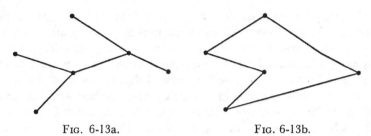

FIG. 6-13a.   FIG. 6-13b.

The tree provides a useful device for the solution of the communication network problem. Consider, for example, that it is necessary to construct a communication network among a large number of cities so that each city is linked either directly or indirectly with every other. If the cost C_{ij} of establishing a link between the ith and the jth city is known, it is possible to compare all of the proposed solutions. In the network portrayed in Figure 6-14 each node represents a city and the value associated with each arc indicates the cost of linking two cities.

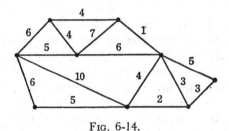

FIG. 6-14.

If it is now our purpose to find the least expensive network connecting all the nodes, we at once realize that the solution must be in the form of a tree, for if a network contains a loop it is always possible to eliminate one of the arcs without isolating a node from the rest of the network. Thus, the problem becomes one of determining the optimal tree. However, this fact alone does not greatly simplify the problem, for in a large network there is a correspondingly large number of possible trees.

Consequently, to solve a problem of this nature, we begin by seeking the link having the lowest cost. If there is more than one link with the same lowest cost, we select one arbitrarily. This link will be part of the optimal tree. From the remaining links we now choose the one with the lowest cost that does not form a loop with the arcs already selected. This procedure is continued until all the cities are connected. The resulting tree is the optimal solution to the network problem. Applying this method to the example portrayed in Figure 6-14, we find that 27 is the lowest possible cost and that the optimal tree is that one illustrated in Figure 6-15. NOTE: The numbers attached to the arcs in this figure indicate their order of selection, not the costs of the links (which are shown in Figure 6-14). This device is practically

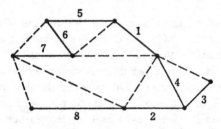

Fig. 6-15.

unique in its simplicity. Unfortunately, comparable methods have not yet been derived for many of the remaining optimal network problems. For example, if we wish to find the least expensive route for a travelling salesman who begins and ends his tour at the same city, the values affixed to the arcs then indicate travelling expenses between two nodes and the solution network will be in the form of a loop. However, there is at present no convenient method for solving this type of problem, which, when the number of cities involved is large, can be extremely complex.

Chapter 7

TRANSFORMATIONS AND TRANSFORMS

The words *transformation* and *transform*, to mention them in the order in which they are treated in this chapter, have a number of uses in mathematics. In fact, the word *transformation* is essentially synonymous with *function*, *operator*, *correspondence*, etc. However, partly for historical reasons and partly for definiteness, the different words are used in different settings, which reduces somewhat the use of the word transformation.

One of its general uses is in connection with changes in coordinate systems. Such changes may be made in the same system, as by translation or rotation of the axes of a rectangular coordinate system; or they may involve a change in the coordinate system itself, e.g., change from rectangular coordinates to spherical coordinates, cylindrical coordinates, or to any of the numerous other kinds of coordinate systems.

Such changes in system naturally change an equation which is expressed in the system. In fact, such changes in equations are one of the reasons why changes in systems are often made, for as is shown in this chapter, an equation can often be simplified by such changes, and therefore be made more useful in computation, and more readily adapted to other mathematical operations.

The word *transform* also has a number of specific present-day uses. Its application to operations with matrices can be shown to follow from the methods of Chapter 2. For if A, B, and X are three matrices then $B = X^{-1}AX$ is the transform of A by X (X being non-singular).

There is also an application of the word *transform* to integral equations since in the equation

$$f(x) = \int K(x, y)F(x)dx \qquad (7\text{-}1)$$

$f(y)$ is called a transform of $F(x)$. This usage gives rise to a number of special kinds of integral transforms. One of them, the Laplace trans-

225

form, has become important in modern applied mathematics, chiefly because of its use in the analysis of feedback and control systems, which is facilitated by reference to tabulated values of the transforms and their corresponding functions, which has the effect of changing problems in analysis into algebraic operations. For that reason the Laplace transform is treated at some length in this chapter.

74. Coordinate Systems. When describing the location or nature of a point, line or surface in three-dimensional space, one type of co-ordinate system may have a distinct advantage over others. Consequently, it is frequently useful to know how to translate from one co-ordinate system to another. Furthermore, some problems may be greatly simplified by moving the origin or reorienting the axes.

We begin by examing the most familiar and elementary of all co-ordinate systems, i.e., the Cartesian, a rectangular coordinate system.

75. Cartesian Coordinates. In three-dimensional space, three reference planes can be used to locate points by giving the distance of the points from each of the reference planes along a line parallel to the intersection of the other two. If the planes are mutually perpendicular,

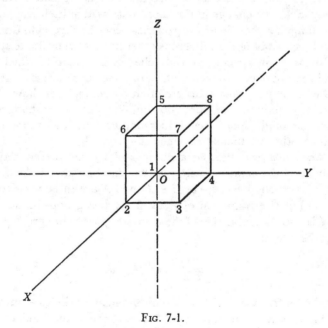

FIG. 7-1.

these distances are called the rectangular Cartesian coordinates of the point in space.   The three intersections of these three planes are called the axes of coordinates and are usually labelled the X-axis, Y-axis and Z-axis.   Their common point is called the origin.   The three co-ordinate planes divide space into eight compartments, called octants. The octant containing the three positive axes as edges is called the first octant.   The other octants are numbered 2 through 8; 2, 3, and 4 are determined by progressing counterclockwise around the positive Z-axis. Then the quadrant vertically beneath the first quadrant is labelled 5 and the remaining quadrants 6, 7, and 8 are labelled by proceeding counterclockwise around the negative Z-axis.

The Cartesian coordinate system is illustrated in Figure 7-1. Note that there are three mutually perpendicular axes and that the part of each axis which is represented by an unbroken line is positive while the part represented by a broken line is negative.   Three of the six faces of the block appearing in Figure 7-1 are contained in the planes which determine the first octant.   If the dimensions of the block's base are 2 units by 2 units and the altitude is 3 units, it may be said to have the following coordinates for each of its vertices:

Vertex	X	Y	Z
1	0	0	0
2	2	0	0
3	2	2	0
4	0	2	0
5	0	0	3
6	2	0	3
7	2	2	3
8	0	2	3

It is easy to see that any three-dimensional figure may be described in terms of Cartesian coordinates.   However, if the restriction is made that $z = 0$, then the base of operations is reduced to 2-dimensional space or, more explicitly, to the XY-plane, i.e., that plane defined by the X-axis and the Y-axis.   If we impose the additional restriction that $y = 0$, then we are left with only the X-axis.   Finally if $x = y = z = 0$, the only remaining point which satisfies the criterion is the origin itself.

By expressing one coordinate as dependent upon another or upon the other two, we are able to express many geometrical figures in terms of a locus based on Cartesian coordinates. For example, a line in 3-dimensional space may be represented by the general equation

$$\frac{x - x_1}{A} = \frac{y - y_1}{B} = \frac{z - z_1}{C} \tag{7-2}$$

where (x_1, y_1, z_1) are the coordinates of a known point on the line and A, B, and C are constants. From this equation it is possible to deduce the more familiar equation for a line in 2-dimensional space, i.e., in the XY plane. Letting $z = z_1 = 0$, we obtain

$$\frac{x - x_1}{A} = \frac{y - y_1}{B}$$

$$\frac{y - y_1}{x - x_1} = \frac{B}{A}$$

$$y - y_1 = \left(\frac{B}{A}\right)(x - x_1)$$

$$y = \left(\frac{B}{A}\right)x + \left(y_1 - \frac{B}{A}x_1\right) \tag{7-3}$$

which reduces to the convenient and more familiar form

$$y = mx + b \tag{7-4}$$

likewise, a plane in 3-dimensional space can be represented by the formula

$$A(x - x_1) + B(y - y_1) + C(z - z_1) = 0$$

or

$$Ax + By + Cz = D \tag{7-5}$$

where D is a constant equal to $Ax_1 + By_1 + Cz_1$.

Similarly, other two- and three-dimensional geometrical figures are represented by expressions in Cartesian coordinates.

One frequently useful way of comparing coordinate systems is to note what types of figures are obtained by holding one coordinate as a constant value and letting the other two assume all possible values. In the case of Cartesian coordinates, if we let $z = c = $ constant and let x and y assume all possible values, we find that the locus of points is a plane parallel to the XY-plane and a perpendicular distance of c away from it. Similarly, if we hold y constant and let x and z range over all possible values, the resulting locus of points is a plane parallel to the XZ-plane and a perpendicular distance away from it equal to the constant value assigned to y.   Finally, if x is assigned a constant value and y and z are permitted to assume all possible values, the resulting locus of points is a plane parallel to the YZ-plane and at a perpendicular distance from it equal to the constant value of x.

76. Spherical Polar Coordinates.   In place of the perpendicular axes of the Cartesian system, spherical polar coordinates are based upon two angular measurements and one distance measurement.   Whereas the three-dimensional Cartesian system is based on the measurements x, y, and z, which give the perpendicular distances from the YZ-, XZ-, and XY-planes respectively, the spherical polar system is based on r, θ and ϕ.   As is illustrated in Figure 7-2, r is the distance from the origin

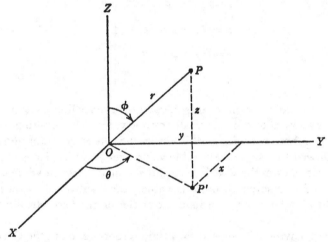

Fɪɢ. 7-2.

to the point P.   Similarly the angle θ is the angle between OX and OP', where P' is the projection* of P on the XY-plane and ϕ is the angle between OZ and OP.

It is evident that the variables in the spherical polar coordinate system are subject to the following restrictions.

$$\left.\begin{array}{c} r \geqslant 0 \\ 0° \leqslant \theta < 360° \\ 0° \leqslant \phi \leqslant 180° \end{array}\right\} \tag{7-6}$$

*The projection of a point in 3-dimensional space upon the XY-plane is simply that point on the XY-plane through which a perpendicular to the plane would pass if drawn through a given point.

Since r, the radius vector, measures the absolute distance between the origin and a point, it cannot assume a value less than zero. Similarly, θ may be thought of, in geographical terms, as a measure of longitude and as such may assume any value in the interval from $0°$ to $360°$. Finally, ϕ may be considered as the colatitude of the given point. Thus, provided that a point is not on the OZ line, it can be represented by a unique set of spherical polar coordinates just as it can be represented by a unique set of Cartesian coordinates.

Noting that $r \sin \phi = OP'$, we obtain the following relationship between the spherical polar and Cartesian systems.

$$\left. \begin{aligned} r^2 &= x^2 + y^2 + z^2 \\ x &= r \sin \phi \cos \theta \\ y &= r \sin \phi \sin \theta \\ z &= r \cos \phi \end{aligned} \right\} \tag{7-7}$$

If the r is held constant in the spherical polar coordinate system and θ and ϕ are allowed to range over all possible values, the resulting locus of points is a sphere of radius r. Similarly, if θ is held constant and r and ϕ are allowed to assume all possible values, the resulting locus of points is a half plane from the Z-axis. And if ϕ is held constant while r and θ are allowed to assume all possible values, the resulting locus of points is a right circular cone with its apex at the origin and axis along the Z-axis.

A sphere having its center at the origin is represented in the Cartesian system as $x^2 + y^2 + z^2 = c^2$, where c is a constant, but in the spherical polar system the corresponding equation is simply $r = c$. Likewise the surface of a cone of revolution around the Z-axis is represented in the Cartesian system as $z^2 = c^2(x^2 + y^2)$. However, in the spherical polar system this equation reduces to $\phi = \text{arc cot } (\pm c)$. Thus it is often more convenient to employ the spherical polar coordinate system when working with certain types of geometrical figures.

Note that this system can also be used to express 2-dimensional figures. If we are simply interested in figures in the XY-plane, then we set ϕ equal to $90°$ which gives $\sin \phi = 1$ and $\cos \phi = 0$. Thus we have

$$x = r \cos \theta \qquad y = r \sin \theta \qquad z = 0 \tag{7-8}$$

Example.   Transform $4x^2 + 4y^2 - z^2 = 0$ into spherical coordinates and describe the surface.

Transposing we have:

$$z^2 = 4(x^2 + y^2)$$

Then substituting:

$$r^2 \cos^2 \phi = 4(r^2 \sin^2 \phi \cos^2 \theta + r^2 \sin^2 \phi \sin^2 \theta)$$

Removing common factors:

$$r^2 \cos^2 \phi = 4r^2 \sin^2 \phi \, [\cos^2 \theta + \sin^2 \theta]$$

Cancelling the common factor r^2 and noting the trigonometric identity $\cos^2 \theta + \sin^2 \theta = 1$:

$$\cos^2 \phi = 4 \sin^2 \phi$$

Dividing both sides by $\sin^2 \phi$:

$$\frac{\cos^2 \phi}{\sin^2 \phi} = 4$$

taking the square root of each side:

$$\frac{\cos \phi}{\sin \phi} = \pm 2$$

Noting the trigonometric identity

$$\frac{\cos \phi}{\sin \phi} = \cot \phi$$

$$\cot \phi = \pm 2$$

taking the inverse trigonometric function*

$$\phi = \text{arc cot} \pm 2$$

$$\phi = 26°34'; \ 153°26'$$

*The inverse trigonometric functions are described in greater detail later in this chapter. $\phi = \text{arc cot} \pm 2$ is read as "ϕ is the angle whose cotangent is equal to plus or minus two."   The inverse trigonometric function defines an angle in terms of one of the trigonometric functions.   If an angle is defined as the inverse of one of its trigonometric functions, it can be redefined by any of them by employing the Pythogorean theorem.   For example, if $\theta = \text{arc tan} \ y/x$, we can visualize a right triangle with the included angle θ.   The side opposite θ is y; the side adjacent to θ is x, and the hypotenuse is therefore, $\sqrt{x^2 + y^2}$.   Thus we have the 3 sides of the triangle and can therefore define θ as the inverse of any of the trigonometric functions.

Thus the figure is a cone of revolution about the Z-axis and is symmetrical with respect to each of the coordinate planes.

Changing coordinates from spherical polar to Cartesian is effected by employing a table of conversions similar to the following:

$$\left.\begin{array}{l} \theta = \arctan y/x \\[2mm] \phi = \arccos \dfrac{z}{\sqrt{x^2+y^2+z^2}} \\[2mm] r = \sqrt{x^2+y^2+z^2} \end{array}\right\} \qquad (7\text{-}9)$$

Example.   Transform $r \sin \phi \tan \phi = 1$ to Cartesian coordinates and describe the resulting figure.

$$r = \sqrt{x^2+y^2+z^2}$$

$$\phi = \arccos \frac{z}{\sqrt{x^2+y^2+z^2}} = \arcsin \frac{\sqrt{x^2+y^2}}{\sqrt{x^2+y^2+z^2}}$$

$$\sin \phi = \sin\left(\arcsin \frac{\sqrt{x^2+y^2}}{\sqrt{x^2+y^2+z^2}}\right) = \frac{\sqrt{x^2+y^2}}{\sqrt{x^2+y^2+z^2}}$$

$$\tan \phi = \frac{\sin \phi}{\cos \phi} = \left(\frac{\sqrt{x^2+y^2}}{\sqrt{x^2+y^2+z^2}}\right)\left(\frac{\sqrt{x^2+y^2+z^2}}{z}\right)$$

$$= \frac{\sqrt{x^2+y^2}}{z}$$

Therefore, $r \sin \phi \tan \phi = \sqrt{x^2+y^2+z^2}\left(\dfrac{\sqrt{x^2+y^2}}{\sqrt{x^2+y^2+z^2}}\right)\left(\dfrac{\sqrt{x^2+y^2}}{z}\right)$

$$= \frac{x^2+y^2}{z}$$

Also, by statement of the problem,

$$r \sin \phi \tan \phi = 1$$

Therefore $z = x^2 + y^2$
which is the equation of a paraboloid.

77. Cylindrical Coordinates.   One way to visualize the cylindrical coordinate system is to consider it the intermediate system between the Cartesian (3 linear measurements) and the spherical polar (1 linear and 2 angular measurements) systems.   The cylindrical coordinate system has two linear and one angular measurements as is illustrated below in Figure 7-3.   Using the cylindrical coordinate system a point is located by assigning values to ρ, θ and z.

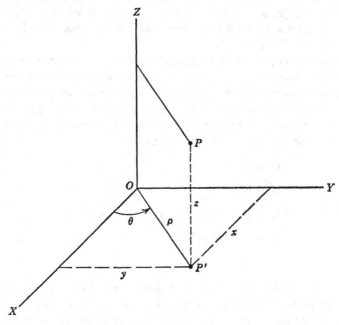

Fig. 7-3.

Two of these symbols are familiar from the coordinate systems already discussed: θ is again the angle between OX and OP' and z is again the perpendicular distance between the point and the XY- (or in this case the ρ-θ) plane.   The new symbol, ρ, is the distance between O, the origin, and P', the projection of P on the XY- or $\rho\theta$-plane.

Thus, we have the following equations relating the Cartesian to the cylindrical coordinate system.

$$\left.\begin{array}{l} x = \rho \cos \theta \\[4pt] y = \rho \sin \theta \\[4pt] z = z \end{array}\right\} \qquad (7\text{-}10)$$

For this system, if we hold ρ constant and allow θ and z to assume all possible values, the resulting locus is a right circular cylinder whose surface is at all points equidistant from the Z-axis. If we hold θ constant and permit ρ and z to assume all possible values, the resulting locus is a half plane from the Z-axis. Finally, if we hold z constant and let ρ and θ assume all possible values, the resulting locus is a plane parallel to the XY-plane.

It is important to note that there is only one point in space having the spherical coordinates (ρ, θ, z) but there is always an infinite number of sets of cylindrical coordinates that will designate any given point. This phenomenon results from the cyclical nature of the trigonometric. function. For any point ρ and z are uniquely determined, but θ may range over an infinite number of values.

$$\theta_0 = \alpha$$

$$\theta_1 = \alpha + 360°$$

$$\theta_2 = \alpha + 720° \qquad \theta_0 = \theta_1 = \theta_2 = \theta_3$$

$$\theta_3 = \alpha + 1080°$$

etc.

By using cylindrical coordinates a right circular cylinder having the Z-axis as its axis of symmetry may be symbolized simply by $\rho = c$, where c is the radius of a circle which is a cross section of the figure.

Example. Transform $x^2 + y^2 = 16$ into cylindrical coordinates. Substituting we have

$$\rho^2 \cos^2 \theta + \rho^2 \sin^2 \theta = 16$$

Factoring:

$$\rho^2 (\cos^2 \theta + \sin^2 \theta) = 16$$

Noting the trigonometric identity $\cos^2 \theta + \sin^2 \theta = 1$ and substituting:

$$\rho^2 = 16$$

which is equivalent to

$$\rho = 4$$

since ρ is a measure of absolute distance without regard to sign.

Just as it is comparatively simple to transform Cartesian coordinates into cylindrical or spherical polar coordinates, so may cylindrical coordinates be transformed with equal ease into spherical coordinates. For such a transformation we rely on the following equations.

$$\left.\begin{array}{cc} \text{Cylindrical} & \text{Spherical} \\[1em] \theta & \theta \\[1em] z & r\cos\phi \\[1em] \rho & r\sin\phi \end{array}\right\} \qquad (7\text{-}11)$$

Thus, from the preceding example, the following three equations are equivalent.

$$x^2 + y^2 = 16 \qquad \text{Cartesian coordinates}$$

$$\rho = 4 \qquad \text{Cylindrical coordinates}$$

$$r\sin\phi = 4 \qquad \text{Spherical polar coordinates}$$

To complete the list of possible transformations between the three systems we have the following conversion tables.

$$\left.\begin{array}{l} \textit{Spherical Polar to Cylindrical Coordinates} \\[1em] \theta = \theta \\[1em] \phi = \text{arc cos } \dfrac{z}{\sqrt{\rho^2 + z^2}} \\[1em] r = \sqrt{\rho^2 + z^2} \end{array}\right\} \qquad (7\text{-}12)$$

$$\left.\begin{array}{l} \textit{Cylindrical to Cartesian Coordinates} \\[1em] z = z \\[1em] \rho = \sqrt{x^2 + y^2} \\[1em] \theta = \text{arc tan } y/x \end{array}\right\} \qquad (7\text{-}13)$$

Example.   First transform $\cot\theta = r\cos\phi$ from spherical into cylindrical coordinates and then transform the resulting equation into Cartesian coordinates.

$$\cot\theta = r\cos\phi \qquad (\text{spherical})$$

From the table relating spherical and cylindrical coordinates $\theta = \theta$, so $\cot \theta = \cot \theta$

From the table, $r = \sqrt{\rho^2 + z^2}$

From the table, $\phi = \text{arc cos} \dfrac{z}{\sqrt{\rho^2 + z^2}}$, so $\cos \phi = \dfrac{z}{\sqrt{\rho^2 + z^2}}$

Substituting these values in the given equation

$$\cot \theta = \sqrt{\rho^2 + z^2}\left(\frac{z}{\sqrt{\rho^2 + z^2}}\right)$$

Thus $\cot \theta = z$ is the transformed equation (cylindrical).

From the table relating cylindrical and Cartesian coordinates

$$\theta = \text{arc tan}\frac{y}{x} = \text{arc cot}\frac{x}{y}, \text{ so } \cot \theta = \frac{x}{y}$$

From the table $z = z$

Thus, substituting these values in the cylindrical equation

$$\frac{x}{y} = z \qquad \text{(Cartesian)}$$

78. The polar coordinate system is essentially a two-dimensional form of either the spherical polar or cylindrical system. Polar coordinates provide an alternate to the Cartesian method of designating points in the XY-plane. It has already been shown how the spherical polar system can be adapted to a two dimensional representation, i.e., ϕ was set equal to 90° and we obtained $x = r \cos \theta$; $y = r \sin \theta$; $z = 0$. Similarly, the same system can be derived from the cylindrical coordinate system by setting $z = 0$. This yields $x = \rho \cos \theta$ and $y = \rho \sin \theta$.

In this case ρ and r are equivalent.

Thus we have the following relationships between Cartesian and polar coordinates.

$$\left.\begin{array}{l} x = \rho \cos \theta \qquad y = \rho \sin \theta \\[2mm] \rho = \sqrt{x^2 + y^2} \quad \theta = \text{arc sin}\dfrac{y}{\sqrt{x^2 + y^2}} = \text{arc cos}\dfrac{x}{\sqrt{x^2 + y^2}} \end{array}\right\}(7\text{-}14)$$

Example 1. Transform $y^2 = 8 + 4x$ into polar coordinates
From (7-14), $y = \rho \sin \theta$ and $x = \rho \cos \theta$
Substituting, $\rho^2 \sin^2 \theta = 8 + 4\rho \cos \theta$.

Example 2.   Transform $\rho = \dfrac{4}{1 - \sin \theta}$ into Cartesian coordinates.

Transposing, $\rho(1 - \sin \theta) = 4$

Substituting from (7-14) $\sqrt{x^2 + y^2}\left(1 - \dfrac{y}{\sqrt{x^2 + y^2}}\right) = 4$

so that $\sqrt{x^2 + y^2} - y = 4$.

79. Hyperbolic Functions.   There is one other system of coordinates to be discussed in this chapter, the bipolar coordinates. Since they are expressed in hyperbolic functions, the latter must first be explained.

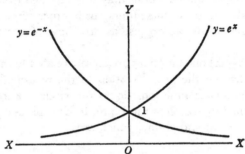

FIG. 7-4.   Graphs of e^x and e^{-x} which define the hyperbolic functions.

The hyperbolic functions are defined as follows (See Fig. 7-4):

$$\left.\begin{aligned}
\sinh x &= \frac{e^x - e^{-x}}{2} \\[6pt]
\cosh x &= \frac{e^x + e^{-x}}{2} \\[6pt]
\tanh x &= \frac{\sinh x}{\cosh x} \\[6pt]
\coth x &= \frac{\cosh x}{\sinh x} = \frac{1}{\tanh x} \\[6pt]
\operatorname{sech} x &= \frac{1}{\cosh x} \\[6pt]
\operatorname{cosech} x &= \frac{1}{\sinh x}
\end{aligned}\right\} \qquad (7\text{-}15)$$

Since $e = \lim\limits_{n \to \infty}\left(1 + \dfrac{1}{n}\right)^n = \lim\limits_{n \to 0}(1 + n)^{1/n}$

$$= 1 + \frac{1}{1!} + \frac{1}{2!} + \frac{1}{3!} + \cdots \frac{1}{n!} \cdots$$

Then, $\left.\begin{aligned} \sinh x &= \frac{e^x - e^{-x}}{2} = x + \frac{x^3}{3!} + \frac{x^5}{5!} + \cdots \\[2mm] \cosh x &= \frac{e^x + e^{-x}}{2} = 1 + \frac{x^2}{2!} + \frac{x^4}{4!} + \cdots \end{aligned}\right\}$ (7-16)

and the other hyperbolic functions may be computed from these values of sin*h* and cos*h*. In using these functions in other computations, their values are found from tables, just as are those of the trigonometric functions.

In fact, the hyperbolic functions are closely related to the trigonometric functions, having the same relation to the rectangular hyperbola that the trigonometric functions do to the circle. Further relationships are obtained by recalling the expressions for series computation of the trigonometric sin and cosine, which are

$$\sin x = x - \frac{x^3}{3!} + \frac{x^5}{5!} - \frac{x^7}{7!} \pm \cdots \cdots$$

$$\cos x = 1 - \frac{x^2}{2!} + \frac{x^4}{4!} - \frac{x^6}{6!} \pm \cdots \cdots$$

In fact, for real values of x, using i for $\sqrt{-1}$,

and $\left.\begin{aligned} \sinh ix &= i \sin x \\ \cosh ix &= \cos x \\ \tanh ix &= i \tan x \\ \sin ix &= i \sinh x \\ \cos ix &= \cosh x \\ \tan ix &= i \tanh x \end{aligned}\right\}$ (7-17)

Moreover, there are relations between hyperbolic functions similar to (but not the same as) those between trigonometric functions, as

$$\left.\begin{array}{l} \cosh^2 x - \sinh^2 x = 1 \\[2mm] 1 - \tanh^2 x = \operatorname{sech}^2 x \\[2mm] \cosh^2 x + \sinh^2 x = 2x \\[2mm] 2 \sinh x \cosh x = \sinh 2x \end{array}\right\} \qquad (7\text{-}18)$$

The inverse hyperbolic functions are as follows:

$$\left.\begin{array}{l} \sinh^{-1} x = \log_e(x + \sqrt{x^2 + 1}) \\[3mm] \cosh^{-1} x = \log_e(x \pm \sqrt{x^2 - 1}) \\[3mm] \tanh^{-1} x = \dfrac{1}{2}\log_e\left(\dfrac{1 + x}{1 - x}\right) \\[4mm] \operatorname{cotanh}^{-1} x = \dfrac{1}{2}\log_e\left(\dfrac{x + 1}{x - 1}\right) \\[4mm] \operatorname{sech}^{-1} x = \log_e\left(\dfrac{1 \pm \sqrt{1 - x^2}}{x}\right) \\[4mm] \operatorname{cosech}^{-1} x = \log_e\left(\dfrac{1 + \sqrt{1 + x^2}}{x}\right) \end{array}\right\} \qquad (7\text{-}19)$$

The inverse hyperbolic functions also have their analogs in plane trigonometry, namely, the inverse trigonometric functions. The inverse of a trigonometric function is the angle which has that function. Inverse trigonometric functions are written either by prefixing the name of the function with the word "arc", or following it by the superior symbol -1, as arc sin θ, $\cos^{-1}\phi$. (Note that the -1 is not an exponent, but means the angle to which the function corresponds). Since the functions of $360° + \theta$, $720° + \theta$, $\cdots$ are the same as those of θ, the values of the inverse functions are taken at the smallest angle to which they apply, unless otherwise specified.

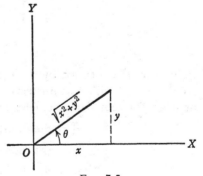

Fig. 7-5.

Consider Figure 7-5. The basic trigonometric functions for the angle θ are the following (the secant and cosecant being reciprocals of the cosine and sine, respectively):

$$\sin \theta = \frac{y}{\sqrt{x^2 + y^2}} \qquad \cos \theta = \frac{x}{\sqrt{x^2 + y^2}}$$

$$\tan \theta = \frac{y}{x} \qquad \cot \theta = \frac{x}{y}$$

Similarly the inverse trigonometric functions are as follows

$$\left. \begin{array}{l} \theta = \text{arc sin } \dfrac{y}{\sqrt{x^2 + y^2}} = \text{arc cos } \dfrac{x}{\sqrt{x^2 + y^2}} \\[2ex] \quad = \text{arc tan } y/x = \text{arc cot } x/y \end{array} \right\} \qquad (7\text{-}20)$$

80. Bipolar Coordinates. This coordinate system is used quite frequently in problems of hydrodynamics and electricity. Consider Figure 7-6 below.

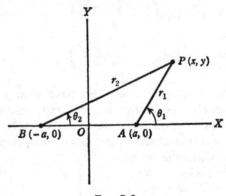

Fɪɢ. 7-6.

Point $P(x, y)$ is defined by finding its orientation with respect to two fixed points A and B symmetrically located on the X-axis a distance $2a$ apart. The four basic elements of this system are

$$r_1 = \text{the distance from } A \text{ to } P$$
$$\theta_1 = \text{the angle } XAP$$
$$r_2 = \text{the distance from } B \text{ to } P$$
$$\theta_2 = \text{the angle } XBP$$

However, in the bipolar coordinate system, these four elements are expressed by two coordinates. These are:

$$\left.\begin{array}{l} \xi = \theta_1 - \theta_2 \\[2mm] \eta = \log_e \dfrac{r_2}{r_1} \end{array}\right\} \tag{7-21}$$

To transform an expression from Cartesian coordinates into bipolar coordinates, we have the following equations for x and y as functions of ξ and η:

$$\left.\begin{array}{l} x = \dfrac{a \sinh \eta}{\cosh \eta - \cos \xi} \\[3mm] y = \dfrac{a \sin \xi}{\cosh \eta - \cos \xi} \end{array}\right\} \tag{7-22}$$

If $\xi =$ constant, $0 \leqslant \xi \leqslant 2\pi$, we have a family of circles with centers on the Y-axis at the point $x = 0$, $y = a \cot \xi$, the radii of the circles being $a \csc \xi$. Each member of this family will pass through the fixed points A and B as shown in Figure 7-7, and will intersect the circles $\eta =$ constant orthogonally. (i.e., at right angles.)

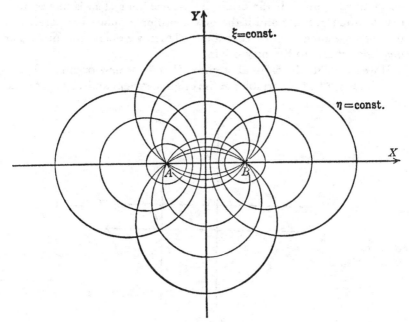

Fig. 7-7.

If η = constant, the result is a second family of circles, having radii of length a cosech η and having centers situated on the X-axis at the points $x = a$ coth η, $y = 0$.   The point A is obtained when $\eta = +\infty$ and B when $\eta = -\infty$.   When $\eta = 0$, the circles degenerate into points on the Y-axis.   The position of a point in the XY-plane is thus fixed when we know in which quadrant it lies and furthermore the constant values of η, ξ of the circles which pass through it.   Since the fixed points A and B (i.e., the X-axis) divide each circle of the set ξ = constant into two segments, we arbitrarily take $\xi = \xi_0 < \pi$ for the arc above the X-axis and $\xi = \xi_0 + \pi$ for all points below this axis.

In order to use these circles as a coordinate systems in space, imagine them to be moved along the Z-axis.   Then as each of the variables in turn is held constant while the remaining two assume all possible values, we obtain the following coordinate surfaces (1) ξ = constant: cylinders with centers on the Y-axis; (2) η = constant: cylinders with centers on the X-axis; (3) z = constant: planes perpendicular to the Z-axis.

81. Translation of Coordinate Axes.   There are many occasions when it is convenient to move the position of the origin relative to a geometrical figure.   If the origin in a Cartesian system is moved to a new location in space and if the new coordinate planes are parallel to the corresponding original coordinate planes, we define this process as *translation of axes* to the new origin.

If we designate the original origin as O and the new origin as O' and if we let x, y, z be the variables in the original system and x', y', z' be the

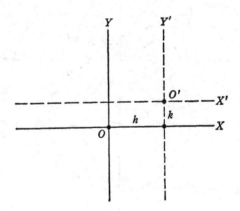

Fig. 7-8.

variables in the translated system, we have the following equations relating the two systems.

$$x = x' + h \qquad y = y' + k \qquad z = z' + l \qquad (7\text{-}23)$$

where (h, k, l) are the coordinates of O' in the original system.

In Figure 7-8 the origin of a two-dimensional Cartesian (x, y) system is translated from $x = 0, y = 0$ to $x = h, y = k$ or $x' = 0, y' = 0$. If, for example, we set $h = 4$ and $k = 3$, we obtain the equations

$$x = x' + 4 \qquad y = y' + 3$$

Thus the point $x = 7$, $y = 13$ in the original system becomes $x' = 3$, $y' = 10$ in the translated system.

Example. Using the same coordinate systems as those described above, i.e.,

$$x = x' + 4 \qquad y = y' + 3$$

Express the equations $x^2 - 8x + y^2 - 6y = 11$ in terms of x' and y'.

We simply substitute $(x' + 4)$ for every x and $(y' + 3)$ for every y in the given equation.   Thus we have

$$(x' + 4)^2 - 8(x' + 4) + (y' + 3)^2 - 6(y' + 3) = 11$$

which gives

$$x'^2 + 8x' + 16 - 8x' - 32 + y'^2 + 6y' + 9 - 6y' - 18 = 11$$

which simplifies to $x'^2 + y'^2 = 36$

This is the equation of a circle of radius 6.   Thus we know that the original equation in x and y is a circle having radius 6 and center at $x = 4, y = 3$.   We know the center of the circle is at the new origin because the equation in x' and y' contains no linear terms.

Similarly, Figure 7-9 shows the translation of the origin in a three-dimensional Cartesian system.   The original axes are shown in solid lines while the translated axes are in dashed lines.

As might be deduced from the preceding example, one of the principal objectives in transformation of coordinates is the simplification of equations in terms of the new coordinates.   For example, any second degree equation in x and y, not containing the product xy, may be simplified to an expression containing at most one linear term, upon a proper translation of axes.

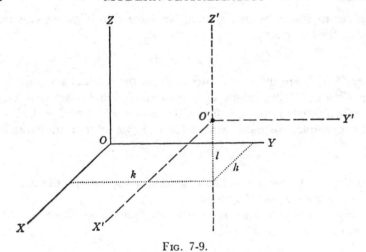

FIG. 7-9.

Thus if we have the equation

$$Ax^2 + Cy^2 + 2Dx + 2Ey + F = 0 \qquad (7\text{-}24)$$

and A and C are not both equal to zero, it is possible to eliminate one or both of the terms Dx, Ey by a suitable translation of axes. Furthermore for any equation of the form given above, we know that (1) if A or C is equal to zero, the equation describes a parabola, two parallel lines, or an imaginary figure; (2) if A and C are of the same sign, the equation describes an ellipse, a single point or an imaginary figure; (3) if A and C are of opposite signs, the equation describes a hyperbola or two intersecting lines. By completing the squares in the given second degree equation, we obtain one of the following general forms.

I. $(y - k)^2 = 2p(x - h)$

    $(x - h)^2 = 2p(y - k)$ $\qquad\qquad (7\text{-}25)$

Parabola with vertex at (h, k) and axis parallel to one of the coordinate axes.

II. $\dfrac{(x - h)^2}{a^2} + \dfrac{(y - k^2)}{b^2} = 1 \qquad (7\text{-}26)$

Ellipse with center at (h, k) and with axes parallel to the coordinate axes.

III. $\left.\begin{array}{l} \dfrac{(x - h)^2}{a^2} - \dfrac{(y - k)^2}{b^2} = 1 \\[12pt] \dfrac{(y - k)^2}{a^2} - \dfrac{(x - h)^2}{b^2} = 1 \end{array}\right\}$ (7-27)

Hyperbola with center at (h, k) and with axes parallel to the coordinate axes.

Thus when we have completed the square of the given equation and obtained one of above forms, we may then translate the origin to (h, k) which will result in the elimination of one or both of the linear terms.

$$x^2 + 4x - 6y + 10 = 0$$

Here we see that $C = 0$. Thus if this equation describes a real geometric figure, we know it must be either a parabola or two parallel lines. Completing a square we find

$$(x^2 + 4x + 4) = 6y - 6$$

$$(x + 2)^2 = 6(y - 1)$$

Thus $h = -2$ and $k = 1$.

To translate the origin to this point, i.e., to the point $(-2, 1)$ we employ the following equations:

$$x = x' - 2 \qquad y = y' + 1$$

Substituting these values for x and y into the original equation we obtain

$$(x' - 2)^2 + 4(x' - 2) - 6(y' + 1) + 10 = 0$$

$$x'^2 - 4x' + 4 + 4x' - 8 - 6y' - 6 + 10 = 0$$

$$x'^2 - 6y' = 0$$

or $$x'^2 = 6y',$$

which is the equation of a parabola.

Example 2. Simplify the equation

$$25x^2 + 9y^2 + 200x - 108y + 499 = 0$$

First group the terms preparatory to completing the squares

$$(25x^2 + 200x) + (9y^2 - 108y) = -499$$

$$25(x^2 + 8x) + 9(y^2 - 12y) = -499$$

Complete the squares

$$25(x^2 + 8x + 16) + 9(y^2 - 12y + 36) = -499 + 400 + 324$$

$$25(x + 4)^2 + 9(y - 6)^2 = 225$$

$$\frac{(x + 4)^2}{9} + \frac{(y - 6)^2}{25} = 1$$

This is the equation of an ellipse with its center at $(-4, 6)$. Thus $h = -4$ and $k = 6$ and we must employ the following equations to translate the axes

$$x = x' - 4 \qquad y = y' + 6$$

Substituting these values into the original equation we obtain:

$$25(x' - 4)^2 + 9(y' + 6)^2 + 200(x' - 4) - 108(y' + 6) + 499 = 0$$

$$25x'^2 - 200x' + 400 + 9y'^2 + 108y' + 324 + 200x' - 800 - 108y'$$
$$- 648 + 499 = 0$$

$$25x'^2 + 9y'^2 = 225$$

$$\frac{x'^2}{9} + \frac{y'^2}{25} = 1 \text{ which is the simplified equation sought.}$$

Similarly, by employing three-dimensional analytic geometry, it is possible to simplify higher order equations. For example, the general equation for a sphere is

$$x^2 + y^2 + z^2 + Ax + By + Cz + D = 0 \tag{7-28}$$

However, by completing the squares, an equation of this kind can be rereduced to the form

$$(x - h)^2 + (y - k)^2 + (z - l)^2 = r^2$$

where (h, k, l) are the coordinates of the center of the sphere in Cartesian space and r is the radius of the sphere. By translating the origin to (h, k, l), the linear terms are eliminated and the equation for the sphere becomes

$$x'^2 + y'^2 + z'^2 = r^2$$

Example. Simplify the equation

$$z^2 + y^2 + z^2 - 6x - 2y + 10z - 1 = 0$$

completing the squares, we obtain

$$(x^2 - 6x + 9) + (y^2 - 2y + 1) + (z^2 + 10z + 25)$$
$$= 1 + 9 + 1 + 25$$
$$(x - 3)^2 + (y - 1)^2 + (z + 5)^2 = 36$$

Thus we have a sphere with center at $(3, 1, -5)$ and radius 6. To translate the origin to $(3, 1, -5)$, we must employ the following equations:

$$x = x' + 3; \quad y = y' + 1; \quad z = z' - 5$$

Substituting these values into the original equation we obtain:

$$x'^2 + y'^2 + z'^2 = 36$$

Furthermore, for *any* equation of the second degree in two variables x, y, i.e., in the form

$$Ax^2 + 2Bxy + Cy^2 + 2Dx + 2Ey + F = 0 \qquad (7\text{-}29)$$

if $B^2 - AC \neq 0$, the locus so described has a unique center at (h, k), and upon transforming the origin to this point we obtain an equation having no linear terms, i.e., in the form

$$Ax'^2 + 2Bx'y' + Cy'^2 + F' = 0$$

The coordinates of the center (h, k) can be found from the original equation by using the following formulae:

$$h = \frac{BE - CD}{AC - B^2} \qquad k = \frac{BD - AE}{AC - B^2}$$

Example. Simplify the following equation by removing the linear terms

$$5x^2 + 4xy + y^2 + 8x + 2y - 4 = 0$$
$$A = 5; \quad B = 2; \quad C = 1; \quad D = 4; \quad E = 1$$
$$B^2 - AC = 4 - 5 = -1 \neq 0$$

Therefore, this equation has a unique center at (h, k) and can be simplified. Solving for h and k, we obtain

$$h = \frac{(2)(1) - (1)(4)}{(5)(1) - (2)^2} = \frac{2 - 4}{5 - 4} = \frac{-2}{1} = -2$$

$$k = \frac{(2)(4) - (5)(1)}{(5)(1) - (2)^2} = \frac{8 - 5}{5 - 4} = \frac{3}{1} = 3$$

Thus we have

$$x = x' - 2 \qquad y = y' + 3$$

Substituting these values into the original equation, we obtain:

$$5(x'^2 - 4x' + 4) + 4(x' - 2)(y' + 3) + (y'^2 + 6y' + 9)$$
$$+ 8(x' - 2) + 2(y' + 3) - 4 = 0$$

Clearing parentheses,

$$5x'^2 - 20x' + 20 + 4x'y' - 8y' + 12x' - 24 + y'^2 + 6y' + 9$$
$$+ 8x' - 16 + 2y' + 6 - 4 = 0$$

Collecting terms we have

$$5x'^2 + 4x'y' + y'^2 = 9$$

82. Rotation of Coordinate Axes. Another useful operation which can be conducted on a coordinate system is called *rotation of axes* and involves the rotation of the coordinate axes about the origin to give a new system. In a two-dimensional Cartesian system, the X- and Y-axis can be rotated about the origin through an angle θ to produce a new system with axes X' and Y'. Thus the relationship between the orig-

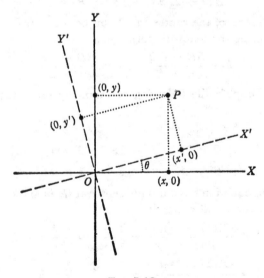

Fig. 7-10.

inal coordinates (x, y) of a point, P, and the new coordinates (x', y') of the same point is given by the following sets of equations:

$$\left.\begin{array}{l} x = x' \cos \theta - y' \sin \theta \\[4pt] y = x' \sin \theta + y' \cos \theta \\[10pt] x' = \quad x \cos \theta + y \sin \theta \\[4pt] y' = -x \sin \theta + y \cos \theta \end{array}\right\} \qquad (7\text{-}30)$$

Such a rotation of axes is illustrated in Figure 7-10.

Example. Find the new equation for $8x^2 + 4xy + 5y^2 = 9$, if the co-ordinate axes are rotated through an angle $\theta = $ arc tan $\frac{1}{2}$.

Using the Pythagorean theorem, we find that for a right triangle containing the angle θ and having the side opposite to θ equal to 1 and the side adjacent to θ equal to 2, the hypotenuse of the triangle is

$$\sqrt{1^2 + 2^2} = \sqrt{5}$$

Thus we have

$$\sin \theta = \sqrt{\frac{1}{5}} \qquad \cos \theta = \sqrt{\frac{2}{5}}$$

Consequently, we obtain:

$$x = (2/\sqrt{5})x' - (1/\sqrt{5})y' = 1/\sqrt{5}(2x' - y')$$
$$y = (1/\sqrt{5})x' + (2/\sqrt{5})y' = 1/\sqrt{5}(x' + 2y')$$

Substituting these values into the original equation, we have

$$\frac{8}{5}(4x'^2 - 4x'y' + y'^2) + \frac{4}{5}(2x'^2 + 3x'y' - 2y'^2)$$
$$+ (x'^2 + 4x'y' + 4y'^2) = 9$$

which reduces to

$$9x'^2 + 4y'^2 = 9$$

Likewise, the axes of a three-dimensional Cartesian coordinate system may be rotated but this is a type of rotation that is more complex than that encountered in the two-dimensional system. Consider the line OP in Figure 7-11. The direction of this line may be defined by determining the values of α_1, α_2, and α_3. The angle between OX and OP is symbolized by α_1, the angle between OY and OP by α_2, and the angle between OZ and OP by α_3. If we let r represent the length of OP we obtain

$$\cos \alpha_1 = \frac{x}{r}, \quad \cos \alpha_2 = \frac{y}{r}, \quad \cos \alpha_3 = \frac{z}{r} \qquad (7\text{-}31)$$

These three cosines are called the direction cosines of the line OP.

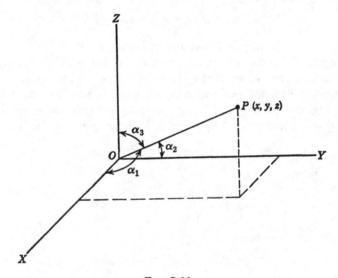

FIG. 7-11.

Now consider a three-dimensional system of Cartesian coordinates with axes OX, OY and OZ and a rotated set of axes OX', OY' and OZ'. Then OX' makes an angle of α_1 with OX, β_1 with OY and λ_1 with OZ. Similarly OY' is related to the original system of axes by angles α_2, β_2, and λ_2, and OZ' is related by angles α_3, β_3, and λ_3. The variables x, y and z of the original system may be related to those of the rotated system (x', y', z') through the direction cosines of these angles. Thus, we have the following equations.

$$\left. \begin{aligned} x &= x' \cos \alpha_1 + y' \cos \alpha_2 + z' \cos \alpha_3 \\ y &= x' \cos \beta_1 + y' \cos \beta_2 + z' \cos \beta_3 \\ z &= x' \cos \lambda_1 + y' \cos \lambda_2 + z' \cos \lambda_3 \end{aligned} \right\} \qquad (7\text{-}32)$$

Just as we saw in the preceding section that, under certain conditions, the linear terms could be removed from an equation of the second degree in x and y by an appropriate translation of axes, we shall now see that, for the same type of equation, it is possible to eliminate the xy term by a suitable rotation of axes.

For an equation of the type that follows, where $B \neq 0$, it is possible to rotate the coordinate axes through an angle θ to eliminate the xy terms.

$$Ax^2 + 2Bxy + Cy^2 + 2Dx + 2Ey + F = 0 \qquad (7\text{-}33)$$

The angle that should be used to effect this transformation may be found by employing the following equation

$$\cot 2\theta = \frac{A - C}{2B} \qquad (7\text{-}34)$$

To derive $\sin \theta$ and $\cos \theta$, once 2θ has been determined, we recall the following trigonometric identities.

$$\left. \begin{array}{l} \cos \theta = \sqrt{\dfrac{1 + \cos 2\theta}{2}} \\[3mm] \sin \theta = \sqrt{\dfrac{1 - \cos 2\theta}{2}} \end{array} \right\} \qquad (7\text{-}35)$$

We need only consider the positive roots in the above expressions since we may assume that when $\cot 2\theta \geqslant 0$, then $0° < 2\theta \leqslant 90°$ and when $\cot 2\theta < 0$, then $90° < 2\theta < 180°$. Consequently, $0° < \theta < 90°$.

Example. Remove the xy term in the equation $4x^2 - 4xy + y^2 = 45$ by rotating the coordinate axes.

Thus we have

$$A = 4 \qquad B = -2 \qquad C = 1$$

$$\cot 2\theta = \frac{A - C}{2B} = \frac{4 - 1}{-4} = -\frac{3}{4}$$

Moreover, from a table of trigonometric functions, the angle having a cotangent of .75 (which is $36°52+'$) has a cosine of .60 or $\frac{3}{5}$. Therefore since the value of $\cot 2\theta$ found is $-.75$, we substitute $-\frac{3}{5}$ for $\cos 2\theta$.

$$\cos \theta = \sqrt{\frac{1 + \cos 2\theta}{2}} = \sqrt{\frac{1 - 3/5}{2}} = \sqrt{1/5}$$

$$\sin \theta = \sqrt{\frac{1 - \cos 2\theta}{2}} = \sqrt{\frac{1 + 3/5}{2}} = 2\sqrt{1/5}$$

Substituting these values in equations (7-30) on Page 249 we have

$$x = \sqrt{\tfrac{1}{5}}\,x' - 2\sqrt{\tfrac{1}{5}}\,y' = \sqrt{\tfrac{1}{5}}\,(x' - 2y')$$

$$y = 2\sqrt{\tfrac{1}{5}}\,x' + \sqrt{\tfrac{1}{5}}\,y' = \sqrt{\tfrac{1}{5}}\,(2x' + y')$$

Substituting these values into the given equation of this problem we obtain

$$4/5(x'^2 - 4x'y' + 4y'^2) - 4/5(2x'^2 - 3x'y' - 2y'^2)$$
$$+ 1/5(4x'^2 + 4x'y' + y'^2) = 45$$

Multiplying both sides of this equation by 5 and removing the parentheses, we have

$$4x'^2 - 16x'y' + 16y'^2 - 8x'^2 + 12x'y' + 8y'^2 + 4x'^2 + 4x'y' + y'^2 = 225$$

Collecting terms:

$$25y'^2 = 225$$

$$y'^2 = 9$$

Thus we see that the original equation defines a pair of parallel lines $y' = \pm 3$ or returning to the notation of the original coordinate system $y = 2x \pm \sqrt{5}$.

It follows from the discussion and examples in this chapter that if the equation

$$Ax^2 + 2Bxy + Cy^2 + 2Dx + 2Ey + F = 0$$

has a real locus, it can be expressed in a simplified form by adopting one of the following procedures. If $B^2 - AC \neq 0$, we first find the center and remove the linear terms by translating the axes to this point. Then to eliminate any xy terms, we rotate the system through the appropriate angle. If on the other hand, $B^2 - AC = 0$, we first rotate the axes to eliminate any xy terms, and then, after completing a square, we may translate the axes to obtain an equation with no more than one linear term.

Example. Simplify the following equation.

$$33x^2 - 8xy + 18y^2 - 116x - 56y + 104 = 0$$

Here $A = 33 \quad B = -4 \quad C = 18 \quad D = -58 \quad E = -28 \quad F = 104$

So that $B^2 - AC = 16 - 594 = -578$, which $\neq 0$
We then proceed to find the center.

$$h = \frac{BE - CD}{AC - B^2} = \frac{112 + 1044}{594 - 16} = \frac{1156}{578} = 2$$

$$k = \frac{BD - AE}{AC - B^2} = \frac{232 + 924}{578} = \frac{1156}{578} = 2$$

Giving the center at $(2, 2)$.
Thus we take

$$x = x' + 2 \qquad y = y' + 2$$

Substituting these into the original equation, we finally obtain:

$$33x'^2 - 8x'y' + 18y'^2 - 68 = 0$$

Next we rotate the axes to eliminate the xy term. In our new equation, we have

$$A = 33 \qquad B = -4 \qquad C = 18$$

$$\cot 2\theta = \frac{A - C}{2B} = \frac{33 - 18}{-8} = \frac{15}{-8}$$

This angle, i.e., 2θ, is shown in Figure 7-12.
From the diagram, we can see that $\cos 2\theta = -15/17$ and we can now find $\cos \theta$ and $\sin \theta$.

$$\cos \theta = \sqrt{\frac{1 - 15/17}{2}} = \sqrt{1/17}$$

$$\sin \theta = \sqrt{\frac{1 + 15/17}{2}} = 4\sqrt{1/17}$$

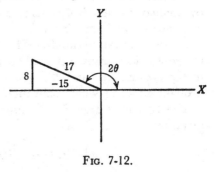

Fig. 7-12.

Thus we have

$$x' = \sqrt{1/17}\,x'' - 4\sqrt{1/17}\,y'' = \sqrt{1/17}(x'' - 4y'')$$

$$y' = 4\sqrt{1/17}\,x'' + \sqrt{1/17}\,y'' = \sqrt{1/17}(4x'' + y'')$$

Substituting these values into the equation in x' and y', we obtain

$$\frac{33}{17}(x''^2 - 8x''y'' + 16y''^2) - \frac{8}{17}(x''^2 - 15x''y'' - 4y''^2)$$

$$+ \frac{18}{17}(16x''^2 + 8x''y'' + y''^2) - 68 = 0$$

which reduces to

$$x''^2 + 2y''^2 = 4$$

This equation, and thus the original equation, describe an ellipse.

83. Laplace Transformations. General Considerations. In view of the broad usefulness of the Laplace transformation in solving time-dependent differential or integro-differential equations, a discussion of these functions belongs properly in a book on modern mathematics. While the transforms themselves are, of course, not modern, their wide use in the study of control systems must certainly be classed with modern applied mathematics.

The behavior of a-c electric circuits is readily expressed in terms of integro-differential equations, as is understood by recalling that the voltage drop across an inductance is a differential function of the current, while the voltage across a capacitance is an integral function of the current. Moreover, many types of non-electrical systems — mechanical, acoustical, hydraulic — lend themselves conveniently to analysis by analogy to electrical systems, so providing a convenient means of generalizing and unifying the analysis of complex systems involving sub-systems of various kinds. This fact, together with the adaptability of the Laplace transform method to the solution of differential and integral equations, and the analysis of discontinuous functions, has brought about its wide application to control problems.

The Laplace transform of a function $f(t)$ is defined as the function given by

$$\mathcal{L}[f(t)] = \int_0^\infty f(t)e^{-st}\, dt = F(s) \tag{7-36}$$

Here the double equation means that the Laplace transform may be written as $\mathcal{L}[f(t)]$ or as $F(s)$, since the transformation process results in a new function $F(s)$ of a variable s instead of the original function f of time, t. The variable s may be real but is usually complex. The transformation is effected by integrating the function $f(t)$ by the use of the factor e^{-st}, where e is the natural logarithmic base (2.71828..). The integral is, as shown, in the improper form, its upper value being considered to increase indefinitely, approaching ∞ as a limit.

84. Evaluating Laplace Transforms. One of the simplest discontinuous functions is the *unit function*, which has only two values, 1 for positive or zero values of t, and 0 for negative values of t; it is often expressed as $u(t)$. Thus for values of $t \geq 0$,

$$\mathcal{L}[u(t)] = \mathcal{L}(1) = \int_0^\infty 1e^{-st}\, dt$$

$$= -\frac{1}{s}e^{-st}\Big]_0^\infty$$

$$= -\frac{1}{s}0 - \left(-\frac{1}{s}1\right)$$

$$= \frac{1}{s} \tag{7-37}$$

The *unit ramp function* is a linear and continuous function of t, thus expressed as $f(t)$ for positive or zero values of t, so that

$$\mathcal{L}[f(t)] = \int_0^\infty te^{-st}\, dt$$

$$= -\frac{te^{-st}}{s}\Big]_0^\infty + \frac{1}{s}\int_0^\infty e^{-st}\, dt$$

$$= 0 + \left[\frac{1}{s}\cdot\frac{-1}{s}e^{-st}\right]_0^\infty$$

$$= \frac{1}{s^2} \tag{7-38}$$

The Laplace transform of a *power of t* is found by a change of the variable of integration, as from t to $w = st$, where s is real and positive

$$\mathcal{L}(t^p) = \int_0^\infty w^p e^{-st}\, dt = \frac{1}{s^{p+1}}\int_0^\infty w^p e^{-w}\, dw \tag{7-39}$$

The last integral at the right is essentially the gamma function, which is given in tables, and which is a generalization of the factorial $n!$, that is defined only for integers and zero. The gamma function may be written as

$$\Gamma(p) = \int_0^\infty w^{p-1} e^{-w}$$

and when $p = n$, the value of the function reduces to $n!$ Therefore we can write for equation (7-40)

$$\mathcal{L}(t^p) = \frac{1}{s^{p+1}} \Gamma(p + 1) \qquad (7\text{-}40)$$

To find the Laplace transform of the *exponential function* e^{-ct} (for positive values of t) where c is a real number:

$$\mathcal{L}(e^{-ct}) = \int_0^\infty e^{-ct} e^{-st} \, dt = \int_0^\infty e^{-(s+c)t} \, dt$$

$$= -\frac{1}{s+c} e^{-st} \Big]_0^\infty$$

$$= \frac{1}{s+c} \qquad (7\text{-}41)$$

To find the Laplace transform of the *trigonometric function* $\sin at$, where a is a real, positive number:

$$\mathcal{L}(\sin at) = \int_0^\infty \sin at \, e^{-st} \, dt.$$

Now in the preceding example, it was found that the evaluation of the Laplace transform of an exponential was quite simple because e^{-st} is also an exponential. So in the present example we replace $\sin at$ by its exponential expression $\dfrac{e^{iat} - e^{-iat}}{2i}$, where $i = \sqrt{-1}$, giving

$$\mathcal{L}(\sin at) = \frac{1}{2i} \int_0^\infty (e^{iat} - e^{-iat}) e^{-st} \, dt$$

$$= \frac{1}{2i} \left(\frac{1}{s - ia} - \frac{1}{s + ia} \right)$$

$$= \frac{a}{s^2 + a^2} \qquad (7\text{-}42)$$

A similar substitution is used to find the Laplace transform of cos at.
Since $\cos at = \dfrac{e^{iat} + e^{-iat}}{2i}$, we have

$$\mathcal{L}(\cos at) = \frac{1}{2i}\int_0^{\infty} (e^{iat} + e^{-iat})e^{-st}\, dt$$

$$= \frac{s}{s^2 + a^2} \tag{7-43}$$

85. Properties of the Laplace Transform. (I) An important
property of the Laplace transform is its linearity, so that the Laplace
transform of the sum of two functions is equal to the sum of the individual Laplace transforms

$$\mathcal{L}[f_1(t) \pm f_2(t)] = F_1(s) \pm F_2(s) \tag{7-44}$$

Here the original functions are denoted by $f_1(t)$ and $f_2(t)$, and their
Laplace transforms by $F_1(s)$ and $F_2(s)$.

The full statement of the linearity (also called superposition) theorem
introduces two constants (a and b below):

$$\mathcal{L}[af_1(t) \pm bf_2(t)] = aF_1(s) + bF_2(s) \tag{7-45}$$

conveying the further information that constant factors of functions
appear unchanged in their Laplace transforms.

Since by equation (7-37) the Laplace transform of $t = 1$ was found
to be $\dfrac{1}{s}$, then it follows directly that the Laplace transform of any
constant is expressed by

$$\mathcal{L}(c) = c\mathcal{L}(1) = \frac{c}{s} \tag{7-46}$$

It also follows from the linearity theorem that the Laplace transform
of any polynomial is equal to the sum of the Laplace transforms of its
terms:

$$\mathcal{L}(a_0 + a_1 t + a_2 t^2 + \cdots a_n t^n) = a_0\mathcal{L}(1) + a_1\mathcal{L}(t)$$
$$+ a_2\mathcal{L}(t^2) + \cdots + a_n\mathcal{L}(t^n) \tag{7-47}$$

The general expression for this series of terms is, therefore

$$\sum_{i}^{n} a_i \frac{\Gamma(i+1)}{s^{i+1}} = \frac{a_0}{s} + \frac{a_1}{s^2} + \frac{2!a_2}{s^3} + \frac{3!a_3}{s^4} \cdots \frac{n!a_n}{s^{n+1}} \tag{7-47a}$$

Note that as derived, Equation (7-47a) is valid only for positive and real values of s. In fact, in all operations with Laplace transforms constant attention must be paid to (1) whether the transform of a function exists at all, and if it does, (2) for what values of the variables it has corresponding real values and for what values it becomes zero or indefinitely great.

II. The real differentiation theorem asserts that if a function and its first derivative have Laplace transforms and if $\mathcal{L}[f(t)] = F(s)$, then

$$\mathcal{L}\frac{df(t)}{dt} = sF(s) - f(t \to 0+) \tag{7-48}$$

which states that the Laplace transform of the first time derivative of a function is equal to the product by s of the Laplace transform of the function itself minus the value of the function as t approaches 0 from the positive side.

III. The real integration theorem, which asserts that if a function of t has a Laplace transform and if $\mathcal{L}[f(t)] = F(s)$ then

$$\mathcal{L}\left[\int f(t)\, dt \right] = \frac{1}{s}\left[F(s) + \int_{0+} f(t)\, dt \right] \tag{7-49}$$

which states that the Laplace transform of the integral of the function is equal to the product by $\frac{1}{s}$ of the Laplace transform of the function plus the value of the integral of the function as t approaches zero from the positive side.

The usefulness of these two theorems is due to the fact, as stated earlier, that so many of the circuit and control equations contain differentials or integrals or both.

IV. There are two translation theorems used with Laplace transforms. The first of these is the time-delay theorem, which is written

$$\mathcal{L}[f(t - t_0)] = e^{-t_0 s}F(s) \tag{7-50}$$

where t_o is real and ≥ 0, provided $f(t) = 0$ for $t < 0$ and, in the case of the $0 +$ transform, $f(t)$ in addition contains no impulse functions at $t = 0$. A direct corrolary of the time-delay theorem is the unit step function. In Equation (7-37) it was found that for values of t of unity, the Laplace transform was $\frac{1}{s}$. The unit step function is $u(t - t_0)$ where t has a value of 1 for all positive and zero values of the function, that is, for all values of t greater than or equal to t_0, while t is zero for all its other values. This function expresses the commencement of a steady-state operation or process of unit value at time t_0.

Equations (7-37) and (7-50) give the unit step function directly, for by (7-37) the Laplace transform of unity is $\frac{1}{s}$, which corresponds to $F(s)$ in (7-50), giving

$$\mathcal{L}u(t - t_0) = e^{-t_0 s} \cdot \frac{1}{s} = \frac{e^{-t_0 s}}{s}$$

V. The time advance theorem is written

$$\mathcal{L}[f(t + t_0)] = e^{t_0 s}F(s) \tag{7-51}$$

Note the change of sign of t_0 on both sides of the equation from those in the time delay theorem (7-50). The same conditions apply: that t_0 is real and ≥ 0, provided $f(t) = 0$ for $t < t_0$ and, in the case of the $0 +$ transform, $f(t)$ in addition contains no impulse functions at $t = t_0$.

VI. The frequency shift theorem is written

$$\mathcal{L}[e^{at}f(t)] = F(s - a) \tag{7-52}$$

where a is any finite constant, real or complex.

VII. The final value theorem asserts that if the function $f(t)$ and its first derivative have Laplace transforms and if $\mathcal{L}[f(t)] = F(s)$, then

$$\lim_{s \to 0} sF(s) = \lim_{t \to \infty} f(t) \tag{7-53}$$

provided that $sF(s)$ is analytic on the imaginary axis and on the right half-plane.

VIII. The corresponding initial value theorem asserts that if a function and its first derivative have Laplace transforms, and if $\mathcal{L}[f(t)] = F(s)$, then

$$\lim_{s \to \infty} sF(x) = \lim_{t \to 0} f(t) \tag{7-54}$$

provided that the $\lim sF(s)$ as s approaches infinity exists.

IX. The last of these Laplace theorems that is widely used in elementary system calculations is the convolution theorem. The convolution of two functions $f_1(t)$ and $f_2(t)$, denoted by $f_1 * f_2 = f_2 * f_1$ is defined by the integral

$$f_1 * f_2 = \int_0^t f_1(t - \tau)f_2(\tau)\, d\tau$$

If $f_1 = t^n$ and $f_2 = t^m$ (n and m, two positive integers) it can be verified by direct calculation that

$$\mathcal{L}\{f_1{}^*f_2\} = F_1(s)F_2(s)$$

where $F_1(s)$ is the Laplace transform of $f_1(t)$ and $F_2(s)$ is the Laplace transform of $f_2(t)$.

thus, if $f_1 = t^2$ and $f_2 = t^3$, then

$$f_1{}^*f_2 = \int_0^t (t-\tau)^2 \cdot \tau^3 \, dt = \frac{t^6}{60}$$

on the other hand $\mathcal{L}(t^2) = \dfrac{2}{s^3}$ and $\mathcal{L}(t^3) = \dfrac{6}{s^4}$

Then

$$\mathcal{L}(t^2) \cdot \mathcal{L}(t^3) = \frac{12}{s^7} = \mathcal{L}\left(\frac{t^6}{60}\right)$$

From this result it follows immediately that if $f_1(t)$ and $f_2(t)$ can be expanded in 2 convergent power series, that

$$(f_1{}^*f_2) = F_1(s)F_2(s). \quad \text{(Convolution theorem)} \qquad (7\text{-}55)$$

In fact the theorem is true in much more general cases. The convolution theorem can be used to derive additional Laplace transforms and also to solve so-called "integral equations" i.e., functional equations where the unknown function appears behind an integral sign.

86. The Inverse Laplace Transform. In solving problems by means of the Laplace transform, the inverse operation of finding functions which correspond to transforms is, of course, as important as the direct one of finding the transforms. For this purpose, it is convenient to use a table of transform pairs, which is given on the following page. The justification for the inverse operation rests, of course, upon the assumption of uniqueness, that for each transform there is only the function. This is true, although its proof is not given here.

The symbol for the operation of inverse (Laplace) transformation is $\mathcal{L}^{-1}$.

There is one kind of inverse transform, however, that is not readily tabulated. It is the inverse transform of a fraction. Since fractions

TABLE 7-1
Table of Laplace Function-Transform Pairs

	$f(t)$	$F(s)$
1.	1	$\dfrac{1}{s}$
2.	c	$\dfrac{c}{s}$
3.	t	$\dfrac{1}{s^2}$
4.	t^p	$\dfrac{\Gamma(p+1)}{s^{p+1}}$
5.	$\sin at$	$\dfrac{a}{s^2+a^2}$
6.	$\cos at$	$\dfrac{s}{s^2+a^2}$
7.	e^{-ct}	$\dfrac{1}{s+c}$
8.	$e^{at}t^n$	$\dfrac{n!}{(s-a)^{n+1}}$
9.	$t \sin at$	$\dfrac{2as}{(s^2+a^2)^2}$
10.	$t \cos at$	$\dfrac{s^2-a^2}{(s^2+a^2)^2}$

occur quite frequently in the problems solved by these methods, the following discussion of one case of finding their inverse is important.

Given a Laplace transform in the form of a rational algebraic fraction

$$F(s) = \frac{A(s)}{B(s)} = \frac{a_m s^m + a_{m-1}s^{m-1} + \cdots + a_1 s + a_0}{b_n s^n + b_{n-1}s^{n-1} + \cdots + b_1 s + b_0} \qquad (7\text{-}56)$$

where the a_i and b_i are all real constants and n and m are positive integers. Where the roots of the denominator are all real or zero and different, a direct method is available to find $\mathcal{L}^{-1}$.

Let the roots of the denominator be $s_1, s_2, \cdots, s_n$. The $F(s)$ takes the form

$$F(s) = \frac{a_m s^m + a_{m-1}s^{m-1} + \cdots + a_1 s + a_0}{(s-s_1)(s-s_2)\ldots\ldots(s-s_n)}$$

This expression may be expressed as the sum of partial fractions

$$F(s) = \frac{k_1}{s - s_1} + \frac{k_2}{s - s_2} + \ldots + \frac{k_n}{s - s_n}$$

where the k_i are coefficients to be determined. The procedure is to multiply both sides of the equation by $(s - s_i)$, which is then equated to 0. Thus expressions for each coefficient are obtained of the general form

$$k_i = \left[(s - s_i) \frac{A(s)}{B(s)} \right]_{s=s_i}$$

where the term $A(s)/B(s)$ represents the transform whose inverse is to be found.

Then using for each coefficient transform-pair #7 in the table of transforms, we have

$$\mathcal{L}^{-1}\left[\frac{k_i}{s - s_i} \right] = k_i e^{s_i t}, \tag{7-57}$$

which can be found for each fraction, so that the inverse transform of $F(s)$ that is sought is the sum of all i values

$$\mathcal{L}^{-1}[F(s)] = \sum^{i} k_i e^{s_i t}, \qquad 0 \leq t \tag{7-58}$$

87. Applications of the Laplace Transform. (a) *Simple Resistance-Capacitance Circuit.* A simple example of the application of the Laplace transform in solving equations is provided by the *RC*-circuit. Let us find, for example, the behavior of the instantaneous current $i(t)$ from the time switch S in Figure 7-13, is closed. The current produced by a given electromotive forces (E) varies inversely as the resistance $\left(i(t) = \frac{E}{R} \right)$ or $(E = i(t)R$, and the integrated current (with respect to time) varies directly as the capacitance $\left(\int i(t)dt = CE \right)$ or $\left(E = \frac{1}{C} \int i(t)dt \right)$ so that the equation for a circuit having both capacitance and resistance is

$$Ri(t) + \frac{1}{C} \int i(t)\, dt = E \tag{7-59}$$

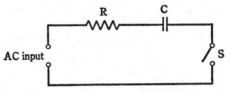

Fig. 7-13.   Resistance-capacitance (RC)
A-C circuit.

Taking the Laplace transform of both sides

$$\mathscr{L}\left[Ri(t) + \mathscr{L}\left[\frac{1}{C}\int i(t)\ dt\right]\right] = \mathscr{L}(E) \tag{7-60}$$

By theorem I, and the definition of the Laplace transform, $\mathscr{L}[Ri(t)] = RI(s)$ where $I(s)$ is the effective value of the current, while by theorem II

$$\mathscr{L}\left[\frac{1}{C}\int i(t)\ dt\right] = \frac{1}{sC}\left[I(s) + \int idt(0+)\right].$$

therefore the equation becomes

$$RI(s) + \frac{1}{sC}\left[I(s) + \int idt(0+)\right] = \frac{E}{s} \tag{7-61}$$

by using transform pair #2 from the table to show that $\mathscr{L}E = \dfrac{E}{s}$

If it is assumed that the initial condition of the circuit is with the switch open then

$$\frac{1}{sC}\int idt(0+) = 0,$$

so that $RI(s) + \dfrac{1}{sC}I(s) = \dfrac{E}{s}$

Transposing $I(s) = \dfrac{\dfrac{E}{s}}{R + \dfrac{1}{sC}}$

$$= \frac{E}{s\left(R + \dfrac{1}{sC}\right)}$$

$$= \frac{E/R}{s + \dfrac{1}{RC}}$$

Substitute P for RC, so that

$$I(s) = \frac{E}{R} \cdot \frac{1}{s + \dfrac{1}{P}} \qquad (7\text{-}62)$$

Taking the inverse transform

$$\mathcal{L}^{-1}[I(s)] = i(t) = \frac{E}{R}\mathcal{L}^{-1}\frac{1}{s + 1/P}$$

Then by inverse transform #7 in the table,

$$i(t) = \frac{E}{R}e^{-t/P} = \frac{E}{R}e^{-t/RC} \qquad (7\text{-}63)$$

which is the instantaneous value of the current for given values of E, R and C, at time t.

(b) *Generalized Resistance-Capacitance Circuit.* In investigating the behavior of control and servo systems, an effective approach is from the point of view of the a-c electrical circuit. Thus, voltage or electromotive force in the electrical system has as its analogs force in a mechanical system, and pressure in an acoustical or hydraulic system. Current in an electrical system is analogous to velocity in a mechanical system or an acoustical system, and to rate of flow in a hydraulic system. Electrical resistance has as its analogues friction in a mechanical system or a hydraulic system, and the real component of acoustic impedance in an acoustical system. These analogies can be extended to capacitance, inductance, impedance, etc.

Because of these analogies between systems, the use of generalized concepts in control problems is particularly valuable. Let us take the

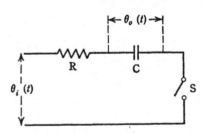

FIG. 7-14. Resistance-capacitance (RC) A-C circuit showing input and output connections.

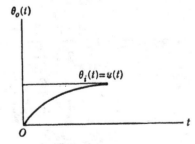

FIG. 7-15. Response of RC circuit to unit input.

first step in this direction by using again the simple resistance-capacitance circuit of the previous application, but denoting the input voltage by $\theta_i(t)$ to differentiate it from the voltage available to an external load by connecting the latter across the condenser, the latter being called the output, $\theta_0(t)$. (See Figure 7-14). Then the electrical equations for these two voltages are

$$\theta_i(t) = Ri(t) + \frac{1}{C}\int i(t)\, dt \tag{7-64}$$

$$\theta_0(t) = \frac{1}{C}\int i(t)\, dt \tag{7-65}$$

Assuming, as in the previous application, that there is no initial charge on C (switch open), then the Laplace transforms of $\theta_i(t)$ and $\theta_0(t)$ are

$$\mathcal{L}[\theta_i(t)] = \theta_i(s) = RI(s) + \frac{1}{sC}I(s) \tag{7-66}$$

$$\mathcal{L}[\theta_0(t)] = \theta_0(s) = \frac{1}{sC}I(s) \tag{7-67}$$

We now generalize these terms: $\theta_i(t)$ is called the driving function of the system; $\theta_0(t)$ is called the response function of the system; $\theta_i(s)$ is called the driving transform of the system; and $\theta_0(s)$ is its response transform. The ratio of the response transform to the driving transform, $\theta_0(s)/\theta_i(s)$ is called the transfer function, and is a characteristic of the system independent of its particular input or output: it is often denoted by $G(s)$.

Then we have from the foregoing equations

$$G(s) = \frac{\theta_0(s)}{\theta_i(s)} = \frac{\frac{1}{sC}I(s)}{RI(s) + \frac{1}{sC}I(s)} = \frac{1}{RCs + 1} \tag{7-68}$$

(c) *Unit Step Driving Function.*

The first Laplace transform evaluated in this section (Equation (7-37) was that of the unit function $u(t)$ for the two values 1 for $t \geqslant 0$, and 0 for $t < 0$. Its Laplace transform was found to be $\dfrac{1}{s}$.

Therefore, if the driving function, $\theta_i(t)$, of a system is of the unit kind, as shown in Figure 7-15, its driving transform is

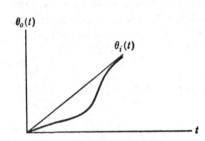

$$\theta_i(s) = \frac{1}{s} \qquad (7\text{-}69)$$

FIG. 7-16. Response of RC circuit to Heaviside unit ramp input.

and its response transform is from Equation (7-68)

$$\theta_0(s) = G(s)\theta_i(s) =$$
$$\frac{1}{RCs + 1} \cdot \frac{1}{s} = \frac{1}{s(RCs + 1)} \qquad (7\text{-}70)$$

Then applying the partial fraction method expressed in Equations (7-57) and (7-58), we have

$$\theta_0(s) = \frac{k_1}{s} + \frac{k_2}{RCs + 1} = \frac{1}{s} + \frac{-RC}{RCs + 1} \qquad (7\text{-}71)$$

Therefore $\theta_0(s) = \dfrac{1}{s} - \dfrac{1}{s - 1/RC}$

Then using transform pairs #1 and #7 in the table,

$$\mathcal{L}^{-1}[\theta_0(s)] = \theta_0(t) = 1 - e^{-t/RC} \qquad (7\text{-}72)$$

Figure 7-15 shows how the value of $\theta_0(t)$ approaches the unit value asymptotically as t increases, as is evident from the equation.

(d) *Unit Ramp Driving Function.*

The second Laplace transform evaluated in this treatment, Equation (7-38), was that of the unit ramp function $f(t)$ for values of $t \geqslant 0$. Its Laplace transform was found to be $\dfrac{1}{s^2}$.

Therefore, if a driving function $\theta_i(t)$ is of the Heaviside unit ramp kind, as shown in Figure 7-16, its driving transform is

$$\theta_i(s) = \frac{1}{s^2}$$

and its response transform is, from Equation (7-70)

$$\theta_0(s) = \frac{1}{s^2(RCs + 1)} \qquad (7\text{-}73)$$

Then applying the partial fraction method expressed in Equations (7-57) and (7-58) we have

$$\theta_0(s) = \frac{k_1}{s^2} + \frac{k_2}{s} + \frac{k_3}{RC+1}$$

$$= \frac{1}{s^2} - \frac{RC}{s} + \frac{(RC)^2}{RCs+1}$$

$$= \frac{1}{s^2} - \frac{RC}{s} + \frac{RC}{s+1/RC}$$

Then by use of transform pairs #1, #2, and #7 in the table, we have

$$\mathcal{L}^{-1}\theta_0(s) = \mathcal{L}^{-1}\left[\frac{1}{s^2} - \frac{RC}{s} + \frac{RC}{s+1/RC}\right]$$

so $\qquad \mathcal{L}^{-1}\theta_0(s) = \theta_0(t) = t - RC(1 - e^{-t/RC})$ \hfill (7-74)

88. Heaviside Operational Calculus. The use of the Laplace transform in solving transient term problems has been developed in this chapter without discussion of the basis from which its use in this way was evolved. Since the Heaviside operational calculus was first used, and in fact, invented for this purpose, a brief account of that system is included. However, its theorems are not included, since they have already been discussed for the Laplace transform.

The behavior of a vibrating system may be analyzed by solving the differential equations of the dynamical system. In other words find the currents or velocities of the elements which when substituted in the differential equations will satisfy the initial and final conditions. The solution of the differential equation may be divided into the steady-state term and the transient term. The operational calculus is of great value in obtaining the transient response of an electrical, mechanical or acoustical system to a suddenly impressed voltage, force or pressure. The general analysis used by Heaviside is applicable to any type of vibrating system whether electrical, mechanical or acoustical. The response of a system to a unit force can be obtained with the Heaviside calculus.

Heaviside's unextended problem is as follows: given a linear dynamical system of n degrees of freedom in a state of equilibrium, find its response when a unit force is applied at any point. The unit function, is defined to be a force which is zero for $t < 0$ and unity for $t \geq 0$.

This is the Heaviside unit step function, which has already been discussed for the Laplace transform.

The response of a dynamical system to a unit force is called the indicial admittance of the system. It is denoted by $A(t)$. $A(t)$ represents the current, linear velocity, angular velocity, or volume current when a unit electromotive force, force, torque or pressure is suddenly applied in a system which was initially at rest.

In the Heaviside calculus the differential equations are reduced to an algebraic form by replacing the operator d/dt by the operator p and the operator $\int dt$ by $1/p$. Tables of operational formulas have been compiled which serve for operational calculus the same purpose that tables of integrals serve the integral calculus. Operational formulas may be modified, divided or combined by various transformation schemes. This is similar to integration by parts or change of variable in the integral calculus

The procedure in the Direct Heaviside Operational Method to be followed in obtaining an operational solution of an ordinary differential equation is as follows: Indicate differentiation with respect to the independent variable by means of the operator p. Indicate integration by means of $1/p$. Manipulate p algebraically and solve for the dependent variable in terms of p. Interpret and evaluate the solution in terms of known operators.

Review of the tables and theorems of Laplace transforms, and the examples of their use, shows the great advantage of the Heaviside method, that is, by use of the impedance concept, one can attack time-varying systems in terms of algebraic rather than differential equations. The correspondence between operational forms and indicial admittance eliminates the necessity for evaluating integration constants. However, the Heaviside method can not be extended to new situations except by a trial-and-error procedure.

While some of the disadvantages of the Heaviside method are overcome by the use of the Laplace transform, that method also has definite limitations. One of the most important of these, which should always be borne in mind, is that the differential equations to which the method is applied must have constant coefficients. (Note that all the theorems state that the functions must have Laplace transforms.)

89. Other Transforms. To show the relationship of some other integral transforms to the Laplace transform, let us start from a general integral equation which has as its kernel a function in two variables $K(x, y)$

$$f(y) = \int K(x, y)F(x)dx$$

Now the function $f(y)$ is an integral transform of $F(x)$. Given $F(x)$, presumably $f(y)$ may be found explicitly. Regarding the equality as an integral equation, however, one wishes to solve for $F(x)$, or invert the transform. Thus, if the transform can be inverted the result will be the solution of the integral equation for the given kernel.

Many special cases have been studied and given special names. The reciprocal relations, written in the form

$$f = C \int KF dx \quad \text{and} \quad F = C' \int K'f dy$$

are shown in the following table for the more familiar integral transforms.

Name	Constants	Integration Limits
1. Fourier............	$C = C' = 1/\sqrt{2\pi}$	$\pm \infty$
2. Sine (cosine)........	$C = C' = \sqrt{2/\pi}$	$\infty, 0$
3. Laplace...........	$C = 1, C' = 1/2\pi i$	$\infty, 0; c \pm i\infty$
4. Mellin............	$C = 1, C' = 1/2\pi i$	$\infty, 0; c \pm i\infty$
5. Hilbert...........	$C = C' = 1/2\pi$	$\pm \pi$

Name	K	K'
1. Fourier............	e^{ixy}	e^{-ixy}
2. Sine (cosine)........	$\sin(\cos) xy$	$\sin(\cos) xy$
3. Laplace...........	e^{-xy}	e^{xy}
4. Mellin............	x^{y-1}	x^{-y}
5. Hilbert...........	$1 + \cot(x - y)/2$	K

Notes for table: (1) The constant factor is sometimes omitted or modified. (3) The constant c is any real number such that $\int_0^\infty e^{-cx}|f(x)|dx$ is bounded. The symbol p is almost invariably used in place of y.

Problems

1. Find the Laplace transform of $15a$ ($a = $ const.).
2. Find the Laplace transform of $2 + 7t + 14t^2$.
3. Find the Laplace transform of $\sin 2t + \cos 3t$.
4. Find the Laplace transform of e^{-2bt}.
5. Find the Laplace transform of $e^{-2bt} t^c$.

6. Find the inverse Laplace transform of $\dfrac{s^2 - b^2}{(s^2 + b^2)^2}$.

7. Find the inverse Laplace transform of $\dfrac{1}{(s - 7)(s - 6)}$.

Chapter 8

NUMERICAL ANALYSIS

90. Nature of Numerical Analysis. The most general definition of numerical analysis is the expression in numbers of an application of mathematics. It means, therefore, the numerical expression of the solution of a mathematical problem. This solution may often be obtained in exact form by the methods of a particular branch of mathematics, such as algebra or calculus. In many instances, however, such exact solutions cannot be found readily (or at all), and in such cases, approximation methods are used. Such methods have been developed extensively by mathematicians for hundreds of years, and have been evaluated for their usefulness in computation. Thus, in choosing a series expansion to approximate a result, an important criterion of the usefulness of a particular series for pencil-and-paper computation is the rapidity with which it converges, which obviously determines the number of terms that must be computed to obtain the result within the desired limits of uncertainty. However, if the computation is to be performed on a computer or other calculating machine, then rapidity of convergence may no longer be the most important criterion of the usefulness of a series in computation. For once the particular program has been established for the machine, it can compute the terms of the series very rapidly, so that other criteria than rapidity of convergence, such as ease of programming, are likely to determine the choice of a series for machine computation. This example is but one out of many in which the development of calculating machines has so radically changed numerical analysis as to give it an important place in a book on modern mathematics. The methods may not be new in principle, but the ones to be chosen for a particular computation, and in fact, the ones most widely used today, have undergone a clearly marked change.

As an example consider the operation of preparing a graph of the polynomial

$$y = x^5 + 7x^4 + 2x^3 + 6x^2 + 2x - 24 \qquad (8\text{-}1)$$

271

for the range $0 \leqslant x \leqslant 20$. Now in computing by pencil and paper methods, one might conveniently use a table of powers of the integers to obtain the various powers and multiply their values by the coefficients in the equations. In working with a calculating machine, however, one would avoid the use of tables or computation of all these powers of x by some process of synthetic division. Such a process is the factoring method shown in successive steps below:

$$y = x^5 + 7x^4 + 2x^3 + 6x^2 + 2x - 24$$

Reversing the polynomial

$$y = -24 + 2x + 6x^2 + 2x^3 + 7x^4 + x^5$$

Factoring the independent variable out of the highest and next highest powers,

$$y = -24 + 2x + 6x^2 + 2x^3 + x(7x^3 + x^4)$$

Factoring the independent variable out of the three highest powers,

$$y = -24 + 2x + 6x^2 + x[2x^2 + x(7x^2 + x^3)]$$

Factoring the independent variable out of the last four terms,

$$y = -24 + 2x + x\{6x + x[2x + x(7x + x^2)]\}$$

Factoring the independent variable out of the last five terms,

$$y = -24 + x|2 + x\{6 + x[2 + x(7 + x)]\}| \qquad (8\text{-}2)$$

Now the computer or calculating machine can be programmed for alternate additions and multiplications with first power values of x. Thus, we have the value of $y = f(x)$ for $x = 5$, or $f(5)$,

$$f(5) = -24 + 5|2 + 5\{6 + 5[2 + 5(7 + 5)]\}| = 7886$$

Note that the word programming is used in two ways. When a problem is to be solved by means of a computer, it is first expressed in a form that is most conveniently adapted to the machine. This step, which might be termed mathematical programming, is illustrated in the foregoing example. When this has been done, the problem must be then coded in the language of the particular computer (unless the machine to be used is a simple manually-operator calculator). This coding is also called programming; however, since the methods in this book are prepared for general use, rather than for a particular computer, it is necessary to restrict the material in this chapter to a discussion of mathematical programming.

91. Difference Tables. In the polynomial given as an example in the preceding section, values of the function were to be computed for integral values of its independent variable. However, in applications of mathematics, it is often necessary to obtain values of such functions for non-integral values of the independent variable. These values may be found, of course, by direct substitution, but if many of them are to be computed, and if a computer is not constantly available, it is often more convenient to use interpolation formulas and a difference table. Another situation in which such methods are used is where the equation expressing the relationship between the independent variable and the function is not known, and we have only a series of corresponding values.

Consider that we have calculated from an equation, or have been given, a succession of values of an independent variable x and its function, or dependent variable, y. Designate these pairs of corresponding values as x_0, y_0; x_1, y_1; x_2, y_2; x_3, y_3; etc., as shown in Table 8-1 below. If the values of x are equidistant, that is, if $x_1 - x_0 = x_2 - x_1 = x_3 - x_2 = x_4 - x_3 \cdots$, then we can form a difference table very simply by arranging in the third column of the table the first order

TABLE 8-1
General Difference Table of Order 5.

x	y	Δ	Δ^2	Δ^3	Δ^4	Δ^5
x_0	y_0					
x_1	y_1	Δy_0				
x_2	y_2	Δy_1	$\Delta^2 y_0$			
x_3	y_3	Δy_2	$\Delta^2 y_1$	$\Delta^3 y_0$		
x_4	y_4	Δy_3	$\Delta^2 y_2$	$\Delta^3 y_1$	$\Delta^4 y_0$	
x_5	y_5	Δy_4	$\Delta^2 y_3$	$\Delta^3 y_2$	$\Delta^4 y_1$	$\Delta^5 y_0$

differences of y, which are $y_1 - y_0, y_2 - y_1, y_3 - y_2 \cdots$. Denoting these first order differences by Δy_i, we would have $y_1 - y_0 = \Delta y_0$; $y_2 - y_1 = \Delta y_1$; $y_3 - y_2 = \Delta y_2$; $\cdots$. We can then obtain second order differences by subtracting successive first order differences, i.e., $\Delta y_1 - \Delta y_0 = \Delta^2 y_0$; $\Delta y_2 - \Delta y_1 = \Delta^2 y_1$; $\Delta y_3 - \Delta y_2 = \Delta^2 y_2 \cdots$ as arranged in Table 8-1. The order of difference can be extended to higher orders subject to the limit that, for a function that is a polynomial of nth degree, the differences of nth order are constant, and those of $(n+1)$th order are zero. This is shown in Table 8-2, which is the difference table for the function (8-1) discussed in the preceding section,

$$y = x^5 + 7x^4 + 2x^3 + 6x^2 + 2x - 24$$

TABLE 8-2
Difference Table for $x^5 + 7x^4 + 2x^3 + 6x^2 + 2x - 24 = y$ for values of x, from 0⁻7

x	y	Δ	Δ^2	Δ^3	Δ^4	Δ^5	Δ^6
0	−24						
1	−6	18					
2	164	170	152				
3	900	736	566	414			
4	3024	2124	1388	822	408		
5	7886	4862	2738	1350	528	120	
6	17484	9598	4736	1998	648	120	0
7	34584	17100	7502	2766	768	120	0

92. Difference Tables Used For Interpolation.   When difference tables are available or may be constructed for equal intervals between values of the independent variable, they are readily used to find values of the function for intermediate values of the independent variable by means of an interpolation formula.   One of the best known of such formulas is Newton's interpolation formula, which may be written in two forms, the first of which is as follows:

$$y = y_a + u\Delta y_a + \frac{u(u-1)}{2!}\Delta^2 y_a + \frac{u(u-1)(u-2)}{3!}\Delta^3 y_a$$
$$+ \ldots + \frac{u(u-1)(u-2)\ldots.(u-n+1)}{n!}\Delta^n y_a \quad (8\text{-}3)$$

where y is any interpolated value of the function to be found for a value of x not given in the table, but within the range of values of the table, y_a, Δy_a, $\Delta^2 y_a \cdots$ are the values of y_a and its differences one place lower in the table than y, $u = \dfrac{x-x_a}{h}$, where x is the value of the independent variable for which interpolated value of y is sought, x_a is the next lower value of x given in the table, and h is the constant difference, $x_1 - x_0 = x_2 - x_1 = \cdots$.

Let us use this interpolation formula to find the value of y corresponding to a value of $x = 1.8372$ in Table 8-2, which is the difference table for the function,

$$y = x^5 + 7x^4 + 2x^3 + 6x^2 + 2x - 24.$$

Thus we have:

$x = 1.8372$

$x_a = 1.0000$

$h = 1$

$u = \dfrac{1.8372 - 1.0000}{1} = .8372$

$$y = (-6) + (.8372)(170) + \frac{(.8372)(-.1628)}{3}(566)$$

$$+ \frac{(.8372)(-.1628)(-1.1628)}{6}(822)$$

$$+ \frac{(.8372)(-.1628)(-1.1628)(-2.1628)}{24}(528)$$

$$+ \frac{(.8372)(-.1628)(-1.1628)(-2.1628)(-3.1628)}{120}(120)$$

$$y = -6.000 + 142.3240 - 38.5718 + 21.7125 - 7.5410 + 1.0841$$
$$= 113.0078$$

This is a close approximation. The exact value of y for $x = 1.8372$ is 113.007795 to six decimal places. In this case, since we knew the equation of y as a function of x (Equation 8-1), it would have been easier to find y by programming the problem by synthetic division (Equation 8-2), and then calculating the solution. However, if we do not have an equation for y as a function of x but only a table of values relating to y and x, it is necessary to employ an interpolation formula such as Newton's.

It is to be noted that the differences used in interpolating by (8-3) lie on a diagonal line in the table. Thus in using it to find y for $x = 1.8372$, the y and Δy terms which appeared in our calculations were y_2, Δy_2, $\Delta^2 y_2$, $\Delta^3 y_2$ and $\Delta^4 y_2$. Suppose, however, that the value of x for which the interpolated value of y was sought had occurred near the bottom of Table 8-2. For example, suppose we had wished to find y for $x = 6.1832$. Then we would have had from the table values for only y_6 and Δy_6 to substitute in the formula, so that a close approximation could not be obtained. In that case, we can avail ourselves of another form of the Newton interpolation formula, which is

$$y = y_a + u\Delta y_{a-1} + \frac{u(u+1)}{2!}\Delta^2 y_{a-2} + \frac{u(u+1)(u+2)}{3!}\Delta^3 y_{a-3}$$
$$+ \ldots + \frac{u(u+1)(u+2)\ldots\ldots(u+n-1)}{n!}\Delta^n y_{a-n} \quad (8\text{-}4)$$

By substituting in this formula for values of $x = 6.1832$, $x_a = 6$, the corresponding values of h, Δy_{6-1}, $\Delta^2 y_{6-2} \cdots$, we have the following figures:

$x = 6.1832$

$x_a = 6.000$

$h = 1.0000$

$u = .1832$

$$y = 17484 + (.1832)(9598) + \frac{(.1832)(1.1832)}{2}(4736)$$
$$+ \frac{(.1832)(1.1832)(2.1832)}{6}(1998)$$
$$+ \frac{(1832)(1.1832)(2.1832)(3.1832)}{24}(648)$$
$$+ \frac{(.1832)(1.1832)(2.1832)(3.1832)(4.1832)}{120}(120)$$

$$y = 17484.0000 + 1758.3536 + 513.2930 + 157.5847 + 40.6729$$
$$+ 6.3016 = 19960.2085$$

This answer is identical with the exact answer as found by direct substitution of 6.1832 for x in Equation (8-2) to an accuracy of three decimal places. In both this and the preceding problem, it is possible to simplify the computational procedure by programming the problem. In this example, we note that the factor (.1832) occurs in each term except the first, that (1.1832) occurs in each term except for the first two, etc., likewise the factor ($\frac{1}{2}$) is a part of the last form terms; the factor ($\frac{1}{3}$) is a part of the last three terms, etc. Thus the problem could be programmed as shown below:

$$y = 17484 + .1832\left|9598 + \frac{1.1832}{2}\left\{4736 + \frac{2.1832}{3}\left[1998 + \frac{3.1832}{4}\right.\right.\right.$$
$$\left.\left.\left.\left(648 + \frac{4.1832}{5}(120)\right)\right]\right\}\right|$$

To calculate the solution from this program, begin by multiplying 4.1832 by 120; divide by 5; add 648; multiply this total by 3.1832; divide this product by 4; add 1998, etc.

This method of interpolation is especially useful for problems involving transcendental functions, which are quite time-consuming to solve directly. Suppose the function e^{-x^2} occurs frequently in an extended series of calculations, which require its evaluation for nonintegral values of x. Then we could save much time by constructing a table of the function for the range of values of x in which we are interested. Thus Table 8-3.

TABLE 8-3

x	$y = e^{-x^2}$	Δ	Δ^2	Δ^3	Δ^4
0.00	1.00000				
0.05	.99750	−.00250			
0.10	.99005	−.00745	−.00495		
0.15	.97775	−.01230	−.00485	+.00010	
0.20	.96079	−.01696	−.00466	+.00019	+.00009
0.25	.93941	−.02138	−.00442	+.00024	+.00005
0.30	.91393	−.02548	−.00410	+.00032	+.00008

Example. Calculate $y = e^{-x^2}$ for $x = 0.2862$. Since this value is near the end of the table, it is better to use the second form of Newton's formula (Equation 8-4) with $x_a = 0.30$ and $u = -0.276$.

$$y = .91393 + (.02548)(.276) + \frac{(.00410)(.276)(.724)}{2}$$
$$- \frac{(.00032)(.276)(.724)(1.724)}{6}$$

$$= 0.91393 + 0.00703 + 0.00041 - 0.00002 = 0.92135$$

This result is correct to the last significant figure shown.

An arrangement of tabulated data, that is somewhat different from the one used above leads to central difference formulas, notably those of Stirling and Bessel. While these converge faster than Newton's formula, this advantage is of no practical importance in machine calculations, although it is often helpful in manual ones, as noted earlier in this chapter.

When the values of x are given for unequal intervals, Newton's formula does not apply, but it is possible to use *divided differences* to obtain interpolated values. The notation used for the designation of divided differences is $\Delta^n[x_i x_j]$ which is read as "the nth order divided difference between x_i and x_j."

The first order divided difference is simply:

$$\Delta[x_i x_j] = \frac{y_i - y_j}{x_i - x_j}$$

which is simply the difference in y divided by the difference in x.

A typical second divided difference is illustrated by the expression

$$\Delta^2[x_0 x_2] = \frac{\Delta[x_0 x_1] - \Delta[x_1 x_2]}{x_0 - x_2}$$

Similarly, a typical third order divided difference is illustrated by the expression:

$$\Delta^3[x_0 x_3] = \frac{\Delta^2[x_0 x_2] - \Delta^2[x_1 x_3]}{x_0 - x_3}$$

Thus, in general, to obtain a divided difference of the nth order, we employ the following formula:

$$\Delta^n[x_i x_{i+n}] = \frac{\Delta^{n-1}[x_i x_{i+n-1}] - \Delta^{n-1}[x_{i+1} x_{i+n}]}{x_i - x_{i+n}} \tag{8-5}$$

In designating a term in a table of divided differences, we employ the notation described above which uses x_i, the independent variable, instead of y_i as in the difference tables previously discussed. The independent variable is used as a basis for the notation in this case because of the unequal interval between the given values of x. Table 8-4 below provides an illustration of the general form of a divided difference table.

TABLE 8-4

x	y	$\Delta[x_i x_j]$	$\Delta^2[x_i x_j]$	$\Delta^3[x_i x_j]$	$\Delta^4[x_i x_j]$	$\Delta^5[x_i x_j]$
x_0	y_0					
x_1	y_1	$\Delta[x_0 x_1]$				
x_2	y_2	$\Delta[x_1 x_2]$	$\Delta^2[x_0 x_1]$			
x_3	y_3	$\Delta[x_2 x_3]$	$\Delta^2[x_1 x_3]$	$\Delta^3[x_0 x_3]$		
x_4	y_4	$\Delta[x_3 x_4]$	$\Delta^2[x_2 x_4]$	$\Delta^3[x_1 x_4]$	$\Delta^4[x_0 x_4]$	
x_5	y_5	$\Delta[x_4 x_5]$	$\Delta^2[x_3 x_5]$	$\Delta^3[x_2 x_5]$	$\Delta^4[x_1 x_5]$	$\Delta^5[x_0 x_5]$

Now consider the particular divided difference table below (Table 8-5). Note that the intervals between the given values of the independent variable are irregular. The first order divided difference, i.e., $\Delta[x_0 x_1]$ was obtained by dividing $(.5978 - .3451)$ by $(2 - 3)$. The remaining divided differences are obtained in a similar fashion by following the general equation given above.

TABLE 8-5

x	y	$\Delta[x_ix_j]$	$\Delta^2[x_ix_j]$	$\Delta^3[x_ix_j]$	$\Delta^4[x_ix_j]$	$\Delta^5[x_ix_j]$
2	.5978					
3	.3451	−.2527				
5	.1890	−.0781	+.0582			
7	.1304	−.0293	+.0122	−.0092		
10	.0891	−.0138	+.0031	−.0013	+.0010	
11	.0806	−.0085	+.0013	−.0003	+.0001	−.0001

There are a number of methods that can be used to find an *approximate* value of y for a value of x not given in the table, but included between the values of x_0 and x_n. One of these methods uses solely the elements in the first diagonal of the divided difference table. To find a value of y corresponding to an x not given in the table, we employ the following formula:

$$y = y_0 + \Delta[x_0x_1](x - x_0) + \Delta^2[x_0x_2](x - x_0)(x - x_1)$$
$$+ \Delta^3[x_0x_3](x - x_0)(x - x_1)(x - x_2) \cdots$$
$$+ \Delta^n[x_0x_n](x - x_0)(x - x_1)(x - x_2) \cdots (x - x_{n-1}) \qquad (8\text{-}6)$$

Example. Suppose we wish to find an approximate value for y when $x = 4.6$, using the data supplied in Table 8-5. We then employ the above interpolation formula and obtain

$$y(x = 4.6) = .5978 + (-.2527)(2.6) + (.0582)(2.6)(1.6)$$
$$+ (-.0092)(2.6)(1.6)(-.4)$$
$$+ (.0010)(2.6)(1.6)(-.4)(-2.4)$$
$$+ (-.0001)(2.6)(1.6)(-.4)(-2.4)(-5.4)$$

After programming the problem to simplify the computation, we obtain

$$y = .2044 \text{ (1.6\% error)}$$

93. Numerical Solution of Algebraic Equations. There is no general method for finding the roots of transcendental equations such as $xe^{-x} = 1$, or $x^2 = \tan x$, though approximate values may always be found by graphical means. However, when more exact results are desired, there are several analytical methods which are often effective. Two of the most useful will be discussed in this section.

One of the classical approaches to the numerical solution of transcendental equations is the Newton-Raphson method. When numerical values for the derivative of $f(x)$ can be found, this method provides a way of determining the real roots of $f(x) = 0$. If x_0 is an approximate value of one of the roots, then a more accurate value of the root (x_1) can be calculated by using the following equation:

$$x_1 = x_0 + \Delta x_0; \qquad \Delta x_0 = -\frac{f(x_0)}{f'(x_0)} \tag{8-7}$$

where $f'(x_0) = \frac{df(x_0)}{dx}$, the derivative.

A subsequent and even more accurate value may be obtained by substituting x_1 in place of x_0.

$$x_2 = x_1 + \Delta x_1; \qquad \Delta x_1 = -\frac{f(x_1)}{f'(x_1)} \tag{8-8}$$

This process may be continued until the desired degree of accuracy is obtained.

To find a first approximation which may be used as a starting point, it is often possible to use graphical methods.

Example. Find the real root of the equation:

$$5x - \sin x - 3 = 0$$

To obtain the first approximation, we begin by transposing the equation to obtain $\sin x$ alone on one side, as:

$$5x - 3 = \sin x$$

Then we set the left hand side of the equation equal to y_1 and the right hand side of the equation equal to y_2, obtaining

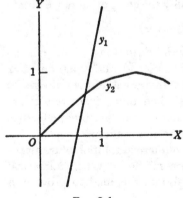

Fig. 8-1.

$$y_1 = 5x - 3$$

$$y_2 = \sin x$$

The graphs of both of these equations are now plotted as shown in Figure 8-1. The value of x at the intersection of the two graphs provides a suitable first approximation. In this case, our first approximation is $x = 0.7$.

We now wish to find a more exact value for x by employing the relations,

$$x_{i+1} = x_i + \Delta x_i; \qquad \Delta x_i = -\frac{f(x_i)}{f'(x_i)} \qquad (8\text{-}9)$$

We have

$$f(x) = 5x - \sin x - 3 \qquad (8\text{-}10)$$

$$f'(x) = \frac{d}{dx}(5x - \sin x - 3) = 5 - \cos x \qquad (8\text{-}11)$$

We may now construct a table (Table 8-6) of values of i, x_i, $f(x_i)$, $f'(x_i)$, $f(x_i)/f'(x_i)$, and Δx_i, by the following calculations:
When $i = 0$, x_i (by graphical approximation above) = 0.7.
Then substituting in Equation (8-10) above,

$$f(x_0) = 5(.7) - \sin (.7) - 3$$

$$= 3.5 - .6442 - 3 = -.1442$$

And substituting in Equation (8-11) above,

$$f'(x_0) = 5 - \cos (.7)$$

$$= 5 - .7648 = 4.2352$$

Then $f(x_0)/f'(x_0) = -.1442/4.2352 = -.0340$
And by Equation (8-9) $\Delta x_0 = -f(x_0)/f'(x_0) = -(-.0340) = .0340$
Whence, also by Equation (8-9),

$$x_1 = x_0 + \Delta x_0 = .7 + .0340 = .7340$$

Then using Equation (8-10)

$$f(x_1) = 5(.7340) - \sin (.7340) - 3$$

$$= 3.6700 - .6698 - 3 = .0002$$

And using Equation (8-11)

$$f'(x_1) = 5 - \cos (.7340)$$

$$= 5 - .7424 = 4.2576$$

Then $f(x_1)/f'(x_1) = \dfrac{.0002}{4.2576} = .00005$

And by Equation (8-9)

$$\Delta x_1 = -\frac{f(x_1)}{f'(x_1)} = -.00005$$

TABLE 8-6

i	x_i	$f(x_i)$	$f'(x_i)$	$f(x_i)/f'(x_i)$	Δx_i
0	.7	−.1442	4.2352	−.0340	.0340
1	.7340	+.0002	4.2576	+.00005	−.00005

In this case only one additional approximation was necessary to obtain x_i accurate to four decimal places.

The Newton-Raphson method may also be used for the solution of ordinary algebraic equations in one variable. Consider the following example.

Example. Find the two real and two imaginary roots of the following equation:

$$x^4 - 10x^3 + 22x^2 - 40x + 72 = 0$$

Our first objective is to find the real roots of the equation. To find feasible first approximation values for x, we construct Table 8-7 by finding $f(x)$ for various integer values of x.

TABLE 8-7

x	0	1	2	3	4	5	6	7	8
$f(x)$	72	45	16	−41	−120	−203	−240	−159	136

From this table we can see that one of the real roots satisfies the inequality $2 < x < 3$ and that the other satisfies the inequality $7 < x < 8$. Thus, we can use $x = 2$ and $x = 7$ for the first approximations.

To simplify the computational process, we program $f(x)$ and $f'(x)$ as follows:

$$f(x) = \{[(x - 10)x + 22]x - 40\}x + 72$$

$$f'(x) = \frac{df(x)}{dx} = 4x^3 - 30x^2 + 44x - 40 = [(4x - 30)x + 44]x - 40$$

Let us first solve for the root that is in the vicinity of 2. Thus, we construct Table 8-8, by the procedure explained earlier in this section.

TABLE 8-8

i	x_i	$f(x_i)$	$f'(x_i)$	$f(x_i)/f'(x_i)$	
0	2.00	+16.00	−40.00	−.40	
1	2.40	−2.35	−51.90	+.05	$x = 2.35$
2	2.35	+.21	−50.36	−.004	

Next we solve for the root that is in the vicinity of 7 by constructing Table 8-9.

TABLE 8-9

i	x_i	$f(x_i)$	$f'(x_i)$	$f(x_i)/f'(x_i)$	
0	7.00	-159.00	$+170.00$	$-.94$	
1	7.94	$+109.95$	$+420.28$	$+.26$	$x = 7.65$
2	7.68	$+11.33$	$+340.39$	$+.03$	
3	7.65	$+1.24$	$+331.71$	$+.003$	

Thus we now have the two real roots of the equation, which are

$$x = 2.35 \qquad x = 7.65$$

Transposing, we obtain:

$$(x - 2.35) = 0 \qquad (x - 7.65) = 0$$

These two expressions are factors of the original equation. If we multiply these two factors together, we have:

$$(x - 2.35)(x - 7.65) = x^2 - 10x + 17.98$$

Rounding off this expression we have:

$$(x - 2.35)(x - 7.65) \doteq x^2 - 10x + 18$$

If we now divide this last expression into the original equation, we obtain:

$$\frac{x^4 - 10x^3 + 22x^2 - 40x + 72}{x^2 - 10x + 18} = x^2 + 4$$

This is then an approximation of the remaining factors in the original equation. Thus we have

$$x^2 + 4 = 0$$

$$x^2 = -4$$

$$x - \pm 2i, \text{ where } i = \sqrt{-1}$$

We have now obtained good approximations of the four roots of the equation. They are as follows:

$$x = 2.35; \quad x = 7.65; \quad x = 2i; \quad x = -2i$$

The method of iteration provides another means for obtaining approximate values for the roots of ordinary equations. If it is possible to rewrite the given equation $f(x) = 0$ in the form

$$x = \phi(x) \tag{8-12}$$

we may substitute an approximate value of x, say x_0, on the right hand side to obtain $x_1 = \phi(x_0)$. This process is then repeated.

$$x_2 = \phi(x_1); \quad x_3 = \phi(x_2); \quad \text{etc.}$$

It is often possible to write $f(x) = 0$ in the form $x = \phi(x)$ in several different ways, in which case it is best to begin with the simplest such arrangement; a few approximations will indicate whether the chosen form is suitable, but if the succeeding values of x do not converge rapidly, one of the alternative functions should be tried. The condition for convergence is found to be that of $\phi'(x)$, the derivative of $\phi(x)$, be less than unity in the vicinity of the desired root. As this derivative becomes smaller, the convergence becomes more rapid.

Example. Solve the equation $x \log_{10}x = 1.5339$ by the method of iteration.

Thus we have

$$x = 1.5334\left(\frac{1}{\log_{10}x}\right)$$

Now let us select $x_0 = 10$ as a first approximation. Then $\log_{10}x_0 = 1$, and the value of x_1 so obtained is

$$x_1 = 1.5334\left(\frac{1}{1}\right) = 1.5334$$

Then
$$x_2 = 1.5334\left(\frac{1}{\log_{10}x_1}\right) = 1.5334\left(\frac{1}{.18564}\right)$$
$$= 8.2573$$

Then
$$x_3 = 1.5334\left(\frac{1}{\log_{10}x_2}\right) = 1.5334\left(\frac{1}{.91684}\right)$$
$$= 1.6726$$

Then
$$x_4 = 1.5334\left(\frac{1}{\log_{10}x_3}\right) = 1.5334\left(\frac{1}{.22339}\right)$$
$$= 6.8640$$

Then
$$x_5 = 1.5334\left(\frac{1}{\log_{10}x_4}\right) = 1.5334\left(\frac{1}{.83658}\right)$$
$$= 1.8329$$

Note that the alternate high and low values obtained by repeated substitutions are slowly converging. This process is quite slow in the case of this equation, as is evident from the following Table 8-10 of values from x_0 to x_{59}.

TABLE 8-10

$x_0 - x_9$	$x_{10} - x_{19}$	$x_{20} - x_{29}$	$x_{30} - x_{39}$
10.0000	4.5528	3.4537	3.2019
1.5334	2.3294	2.8486	3.0340
8.2573	4.1748	3.3730	3.1831
1.6726	2.4708	2.9041	3.0510
6.8640	3.9037	3.3118	3.1656
1.8329	2.5923	2.9483	3.0637
5.8281	3.7065	3.2654	3.1533
2.0031	2.6950	2.9838	3.0742
5.0826	3.5610	3.2295	3.1441
2.1716	2.7799	3.0120	3.0823

$x_{40} - x_{49}$	$x_{50} - x_{59}$		
3.1364	3.1178		
3.0890	3.1047		
3.1305	3.1166		
3.0941	3.1059		
3.1269	3.1154		
3.0972	3.1071	Converging eventually to 3.1110	
3.1231	3.1141		
3.1002	3.1085		
3.1205	3.1135		
3.1028	3.1091		

The convergence of the upper and lower values of x are illustrated by the graph in Figure 8-2. In this figure alternate values of x are joined by two smooth curves which converge toward the value 3.1110. The computation shown above could have been considerably shortened by trying new values for x after the pattern of convergence had been established. One way would have been to select the mean value of two successive approximations and continue the procedure from that value. This process could be repeated at intervals to speed the calculation. An even more accurate way to reduce the computational labor for this particular problem would be to select the mean value of the logarithms for two consecutive values of x and proceed from the corresponding value of x. This operation could also be repeated at intervals.

The method of iteration may also be extended to a system of equations in more than one unknown. Consider the following example:

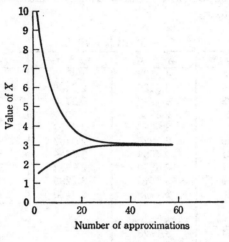

FIG. 8-2.

Example. Solve the following system of equations by the method of iteration.

$$\sin x + y = 1 \tag{8-13}$$

$$\cos y - x = 0 \tag{8-14}$$

First we rewrite the two equations, expressing them in terms of y in the first equation and x in the second.

$$y = 1 - \sin x \tag{8-15}$$

$$x = \cos y \tag{8-16}$$

Now we begin by selecting an arbitrary value for x and substituting it in Equation (8-15). Let us say $x = 0$. Then we obtain $y = 1$ in Equation (8-15). We then substitute this value for y in Equation (8-16) and obtain $x = .5403$. This value is then substituted into Equation (8-15) and a new value is obtained for y, etc. This process is continued until the values of the variables have converged to the desired accuracy. The results of this problem are given below in Table 8-11.

TABLE 8-11

Number of Approximations	0	1	2	3	4	5	6
x	0	.5403	.8843	.9743	.9850	.9860	.9862
y	1	.4859	.2268	.1729	.1668	.1662	.1662

Thus we have:

$$x = .9862$$

$$y = .1662$$

One of the most satisfactory methods of evaluating ordinary systems of linear equations has already been discussed at some length in Chapter 2. The method referred to is, of course, the use of determinants. The procedures described there for the solution of a system of n linear equations in n variables may very properly be classed as one of the most fundamental and valuable tools in the numerical analysis of algebraic equations. Besides that method, and the two treated in this chapter, there are many other methods for the numerical analysis of equations, for which more specialized books should be consulted.

94. Numerical Integration. There are a number of means by which the value of a definite integral may be approximated and choice of a method depends upon the nature of the function to be integrated. For example, if $f(x)$ is known to be continuous over an interval of x from a to b, and either the explicit form of $f(x)$ is unknown or it is such a function that its definite integral cannot be determined conveniently, it becomes necessary to resort to one of the following numerical methods to obtain a satisfactory approximation.

The most elementary approach to this type of problem is known as the *trapezoidal rule*. To use this rule, the values of $f(x)$ between the limits of integration must be positive. If this condition is fulfilled, it is then possible to proceed by dividing the interval (a, b) (where a and b are the limits of integration) into n equal sub-intervals, each having a length of Δx. Ordinates are then erected from the end points of each of the subintervals and their upper ends are joined by straight line segments. The area under the curve $f(x)$ between a and b, which is symbolized by the integral $\int_a^b f(x) \, dx$, can now be approximated by finding the total area contained in the set of trapezoids thus constructed. For each of the ordinates $y_0, y_1, y_2 \cdots y_n$, there is a corresponding abscissa, $x_0, x_1, x_2 \cdots, x_n$ such that $x_0 = a$, $x_1 = a + \Delta x$, $x_2 = a + 2\Delta x$, $\cdots, x_n = a + n\Delta x = b$. The values of x and y are thus related by the formula $y_i = f(x_i)$ where $i = 0, 1, 2, 3, 4, \cdots, n$. We then have a set of trapezoids whose areas are given by the following expressions.

$$0.5(y_0 + y_1)\Delta x$$

$$0.5(y_1 + y_2)\Delta x$$

$$\cdots\cdots\cdots\cdots$$

$$\cdots\cdots\cdots\cdots$$

$$\cdots\cdots\cdots\cdots$$

$$0.5(y_{n-1} + y_n)\Delta x$$

When these terms are added together, we obtain:

$$I_T \doteq (0.5y_0 + y_1 + y_2 + \cdots + y_{n-1} + 0.5y_n)\Delta x, \qquad (8\text{-}17)$$

where I_T = the approximate value of the integral $\int_a^b f(x)dx$ by the trapezoidal rule.

Example. Calculate the approximate value of the integral $\int_0^2 \sqrt{12 - x^3}\, dx$, with $n = 4$.

First we must divide the interval $(0, 2)$ into four subintervals and find the value of $f(x)$ for each of the endpoints, as shown in Table 8-12.

TABLE 8-12

$x_0 = 0.0$	$y_0 = \sqrt{12 - 0} = 3.464$
$x_1 = 0.5$	$y_1 = \sqrt{12 - 0.125} = 3.446$
$x_2 = 1.0$	$y_2 = \sqrt{12 - 1} = 3.317$
$x_3 = 1.5$	$y_3 = \sqrt{12 - 3.375} = 2.937$
$x_4 = 2.0$	$y_4 = \sqrt{12 - 8} = 2.000$
	$\Delta x = 0.5$

Substituting these values in Equation (8-17) for I_T we obtain:

$$I_T = (1.732 + 3.446 + 3.317 + 2.937 + 1.000)(0.5)$$

$$I_T = 6.216$$

Another method, very similar to that of the trapezoidal rule, is known as *Simpson's parabolic rule.* Again it is necessary that $f(x)$ be positive in the interval (a, b) and again the interval (a, b) is divided into n subintervals of uniform magnitude, i.e., Δx. However, in order to apply this rule, n, the number of subintervals, must be an *even* number. The ordinates at the endpoints of each of the subintervals is determined, as was done for the trapezoidal rule, and then the following formula is employed to arrive at an approximate value of the integral.

$$I_s \doteq (y_0 + 4y_1 + 2y_2 + 4y_3 + 2y_4 + \ldots + 4y_{n-1} + y_n)\frac{\Delta x}{3}$$

(8-18)

where I_s is the approximate value of the integral $\int_a^b f(x)dx$ by Simpson's rule.

Example. Using Simpson's rule with $n = 4$, approximate the value of the integral given in the preceding example.
The integral in that example was

$$\int_0^2 \sqrt{12 - x^3}\, dx,$$

and by setting $f(x) = \sqrt{12 - x^3} = y$, we have, as before, the values of y_i for values of x at intervals $\Delta x = 0.5$ from 0 to 2, shown in Table 8-13.

TABLE 8-13

$x_0 = 0.0$	$y_0 = 3.464$
$x_1 = 0.5$	$y_1 = 3.446$
$x_2 = 1.0$	$y_2 = 3.317$
$x_3 = 1.5$	$y_3 = 2.937$
$x_4 = 2.0$	$y_4 = 2.000$
$\Delta x = 0.5$	

Substituting in Simpson's formula (8-18), we have

$$I_s = [3.464 + 4(3.446) + 2(3.317) + 4(2.937) + 2.000]\left(\frac{0.5}{3}\right)$$

$$I_s = [3.464 + 13.784 + 6.634 + 11.748 + 2.000](.167)$$

$$I_s = 6.284$$

Another method, again very similar to the two already described, is known as Weddle's rule. This is the most accurate method of those discussed so far, but it has the disadvantage that n must be six or a multiple of 6. Otherwise the same procedure as that used in the preceding methods is employed and the values are substituted into the following equation.

$$I_w \doteq \int_a^b f(x)dx \doteq (.3\Delta x)[y_0 + 5y_1 + y_2 + 6y_3 + y_4 + 5y_5 + 2y_6 + 5y_7$$
$$+ y_8 + 6y_9 + y_{10} + 5y_{11} + 2y_{12} + \ldots + 2y_{n-6} + 5y_{n-5} + y_{n-4}$$
$$+ 6y_{n-3} + y_{n-2} + 5y_{n-1} + y_n \qquad\qquad (8\text{-}19)$$

Example — Evaluate $\int_0^{1.5} \dfrac{x^3}{e^x - 1}dx$ by Weddle's rule with $n = 6$.

This integral is of real importance in the Debye theory of the heat capacity of solids; it cannot be evaluated in terms of other known functions. Dividing the interval 0 into 1.50 into six equal parts, and substituting the corresponding values of x in the expression to be integrated, we obtain Table 8-14.

TABLE 8-14

$x_0 = 0.00$	$y_0 = 0.0000$
$x_1 = 0.25$	$y_1 = 0.0550$
$x_2 = 0.50$	$y_2 = 0.1927$
$x_3 = 0.75$	$y_3 = 0.3777$
$x_4 = 1.00$	$y_4 = 0.5820$
$x_5 = 1.25$	$y_5 = 0.7843$
$x_n = 1.50$	$y: = 0.9694$

Substituting these values into the equation for I_w, (Weddle's rule) we obtain:

$$I_w = (.3 \cdot 0.25)[0 + 5(0.055) + 0.1927 + 6(0.3777) + 0.5820$$
$$+ 5(0.7843) + 0.9694]$$

$$I_w = (.075)[.0000 + .2750 + .1927 + 2.2662 + .5820 + 3.9215$$
$$+ .9694]$$

$$I_w = (.075)(8.2068) = .6155$$

Of the three methods presented above, Weddle's rule provides the most accurate approximation, but Simpson's rule yields a sufficient degree of accuracy for most problems. The trapezoidal rule is the least exact of the three. Of course, the accuracy of each equation depends upon the number of subintervals constructed for the given interval (a, b) and thus upon the magnitude of each subinterval, i.e., Δx.

If the explicit form of $f(x)$ is known and if it has finite derivatives at the upper and lower limits of the integral or if these derivatives may be determined by numerical methods, the *Euler-Maclaurin* formula may be used to evaluate the integral. Indicating the values of $f(x)$ at $x = a$ and at $x = b$ by y_0 and y_n and the intermediate values by y_1, y_2, y_3, this formula is written for odd r

$$\int_a^b f(x)dx = \Delta x \left[\frac{y_0}{2} + y_1 + y_2 + \ldots + \frac{y_n}{2} \right] - \sum \frac{(\Delta x)^{r+1}}{(r+1)!}$$
$$[B_{r+1} y_n^{(r)} - y_0^{(r)}] \quad (8\text{-}20)$$

where $y_n^{(r)}$ and $y_0^{(r)}$ are the rth derivatives of $f(x)$ at the points b and a. The numerical coefficients B_r are the Bernoulli numbers, which are the numerical values of the coefficients of $x^0/0!$, $x^2/2!$, $x^4/4!$, $\cdots x^{2n}/(2n)!$ in the expansion of $x/(1 - e^{-x})$ or $xe^x/(e^x - 1)$. The first four terms in the series are then

$$1 + (1/2)x + (1/6)x^2/2! - (1/30)x^4/4!$$

The odd terms drop out after the term $\frac{1}{2}x$, and the even terms continue. The first ten Bernoulli numbers are as follows:

$B_0 = 1$	$B_6 = \dfrac{1}{42}$
$B_1 = -\dfrac{1}{2}$	$B_7 = 0$
$B_2 = \dfrac{1}{6}$	$B_8 = -\dfrac{1}{30}$
$B_3 = 0$	$B_9 = 0$
$B_4 = -\dfrac{1}{30}$	$B_{10} = \dfrac{5}{66}$
$B_5 = 0$	

Note that in the Euler-Maclaurin formula given above the first term is equal to I_T, the approximation formula given by the trapezoidal rule. Thus we may write the Euler-Maclaurin formula in the following expanded form:

$$\int_a^b f(x)dx = I_{EM} = I_T - \frac{(\Delta x)^2}{12}[y_n' - y_0'] + \frac{(\Delta x)^4}{720}[y_n''' - y_0''']$$
$$- \frac{(\Delta x)^6}{30240}[y_n^{\mathrm{V}} - y_0^{\mathrm{V}}] + \frac{(\Delta x)^8}{1209600}(y_n^{\mathrm{VII}} - y_0^{\mathrm{VII}}) + \ldots \quad (8\text{-}21)$$

Example. Divide the interval between 1.0 and 2.0 into five equal subintervals and evaluate the integral $I = \int_{1.0}^{2.0} \frac{dx}{x}$ by the Euler-Maclaurin formula. The required values of $y = f(x) = 1/x$ are given in Table 8-15. Note that, for the utilization of this method, only the first two columns in Table 8-15 are necessary. The last four columns represent the difference table corresponding to the values of $f(x)$ for the equal subintervals of x and will be used in the discussion of the next method.

TABLE 8-15

x	$f(x) = y$ $= 1/x$	Δy	$\Delta^2 y$	$\Delta^3 y$	$\Delta^4 y$
1.0	1.000000				
1.2	0.833333	−0.166667			
1.4	0.714286	−0.119047	+0.047620		
1.6	0.625000	−0.089286	+0.029761	−0.017859	
1.8	0.555556	−0.069444	+0.019842	−0.009919	+0.007940
2.0	0.500000	−0.055556	+0.013888	−0.005954	+0.003965

First we find I_T by the trapezoidal method (Equation 8-17)

$$I_T = \Delta x[0.5\,y_0 + y_1 + y_2 + y_3 + y_4 + 0.5\,y_5]$$

$$= (0.2)[(0.5)(.1) + .833333 + .714286 + .625000 + .555556$$
$$+ (0.5)(0.5)]$$

$$= (0.2)[.500000 + .833333 + .714286 + .625000 + .555556$$
$$+ .250000] = 0.695635$$

We also need the derivatives of odd order of $f(x) = y = \frac{1}{x}$. These are as follows:

$$f'(x) = \frac{d\left(\frac{1}{x}\right)}{dx} = -x^{-2}; f'''(x) = \frac{d'''x}{dx'''} = -6x^{-4}; f^V(x) = \frac{d^V x}{dx^V} = -120x^{-6}$$

In this problem $x_0 = 1$ and $x_n = 2$. Thus we have:

$$f'(2) = (-2)^{-2} = -0.25; f'(1) = (-1)^{-2} = -1;$$

$$f'''(2) = (-6)2^{-4} = -0.375; f'''(1) = (-6)1^{-4} = -6;$$

$$f^V(2) = (-120)2^{-6} = -1.875; f^V(1) = (-120)1^{-6} = -120$$

We also calculate

$$\frac{(\Delta x)^2}{12} = \frac{(.2)^2}{12} = .003333$$

$$\frac{(\Delta x)^4}{720} = \frac{(.2)^4}{720} = 2.22 \times 10^{-6}$$

$$\frac{(\Delta x)^6}{30240} = \frac{(.2)^6}{30240} = 2.4 \times 10^{-9} \tag{8-21}$$

Substituting these values into the Euler-Maclaurin formula (8-21), we obtain

$$I_{EM} = 0.695635 - (.003333)(-.25 - 1) + (2.22 \times 10^{-6})$$
$$(-0.375 - 6.) - (2.4 \times 10^{-9})(-1.875 - 120.)$$

$$I_{EM} = 0.695635 - (.003333)(0.75) + (2.22 \times 10^{-6})(5.625)$$
$$- (2.4 \times 10^{-9})(118.1)$$

$$= 0.695635 - 0.002500 + .000012 - (2.8 \times 10^{-7})$$

$$= .693147$$

It can be verified that this result is extremely accurate. This can be checked, because in this case, the integral of the function can be readily found. Recall the integration formula

$$\int_a^b \frac{dx}{x} = \log_e x \bigg]_a^b$$

So for this problem

$$\int_1^2 \frac{dx}{x} = \log_e 2 - \log_e 1$$

$$= .693146 - .00000$$

$$= .693146$$

In the case where the explicit form of $f(x)$ is unknown, we may employ *Gregory's formula*, which is essentially the Euler-Maclaurin formula adapted for use with a difference table instead of the derivatives of the function. Thus for Gregory's formula we have

$$\int_a^b f(x)dx = I_{Gr} = I_T - \frac{\Delta x}{12}(\Delta y_{n-1} - \Delta y_0) - \frac{\Delta x}{24}(\Delta^2 y_{n-2} + \Delta^2 y_0)$$
$$- \frac{19(\Delta x)}{720}(\Delta^3 y_{n-3} - \Delta^3 y_0) \frac{-3(\Delta x)}{160}(\Delta^4 y_{n-4} + \Delta^4 y_0) - \dots. \tag{8-22}$$

where I_T is found by the trapezoidal method (Equation 8-17).

Note that the contents of the parentheses are alternately differences and sums. Additional coefficients of $-(\Delta x)(\Delta^r y_{n-r} \pm \Delta^r y_0)$ may be found by evaluating the definite integral

$$\frac{(-1)^r}{(r+1)!} \int_0^1 x(x-1)(x-2) \ldots (x-r)dx \qquad (8\text{-}23)$$

Example. Evaluate the integral of the preceding example by means of Gregory's formula.

We already have calculated (on page 00) $I_T = 0.695635$ and the required differences are given in Table 8-15. Thus, it only remains to compute the coefficients before the formula can be applied. Thus we have

$$\frac{\Delta x}{12} = \frac{.2}{12} = 0.016667$$

$$\frac{\Delta x}{24} = \frac{.2}{24} = 0.008333$$

$$\frac{19(\Delta x)}{720} = \frac{3.8}{720} = 0.005278$$

$$\frac{3(\Delta x)}{160} = \frac{.6}{160} = 0.003750$$

Hence from Equation (8-22)

$I_{Gr} = 0.695635 - (0.016667)(-0.055556 + 0.166667)$
  $- (0.008333)(0.013888 - 0.047620) - (0.005278)(-0.005954$
  $+ 0.017859) - (0.005278)(-0.005954 + 0.017859)$
  $- (0.003750)(0.003965 + 0.007940)$

  $= 0.695635 - 0.001852 - 0.000513 - 0.000063 - 0.000045$

  $= .693162$

This result is not as precise as that obtained in the preceding example because of the limited number of available differences. However, both of these last two methods may be carried out to any desired degree of accuracy.

95. Numerical Differentiation. In order to determine the numerical derivatives of a function of x at a given point, the slope of the curve of the function may be found by graphical means or the data may be fitted to an empirical equation which is then differentiated. However, there are a number of other excellent means available for this solution of such a problem.

From basic calculus, we know that $\frac{dy}{dx}$ may be expressed by the product $(\partial y/\partial u)(\partial u/\partial x)$. Now recall Newton's interpolation formula from the first part of the chapter (Equation 8-3).

$$y = y_a + u\Delta y_a + \frac{u(u-1)}{2!}\Delta^2 y_a$$
$$+ \frac{u(u-1)(u-2)}{3!}\Delta^3 y_a + \ldots + \frac{u(u-1)(u-2)\ldots(u-n+1)}{n!}\Delta^n y_a$$

where $u = \frac{x - x_a}{h}$.

To obtain the desired equation for the derivative we now differentiate $u = \frac{x - x_a}{h}$ and substitute in

$$\frac{dy}{dx} = (\partial y/\partial u)(\partial u/\partial x)$$

$$\frac{\partial u}{\partial x} = \frac{d}{dx}\left(\frac{x - x_a}{h}\right) = \frac{1}{h}$$

Thus we have

$$\frac{dy}{dx} = \frac{1}{h}\frac{\partial y}{\partial u} = \frac{1}{h}\left[\Delta y_a + \frac{(2u-1)}{2!}\Delta^2 y_a\right.$$
$$\left. + \frac{(3u^2 - 6u + 2)}{3!}\Delta^3 y_a + \frac{4u^3 - 18u^2 + 22u - 6}{4!}\Delta^4 y_a \ldots\right]$$

At the point $x = x_a$, $u = 0$. By differentiating again and evaluating the coefficients in both the first and second derivatives we obtain the following expressions for the first and second derivatives of the function.

$$\left.\begin{array}{l}\left(\dfrac{dy}{dx}\right)_{x=x_a} = \dfrac{1}{h}\left[\Delta y_a - \dfrac{1}{2}\Delta^2 y_a + \dfrac{1}{3}\Delta^3 y_a - \dfrac{1}{4}\Delta^4 y_a \ldots\right] \\[2ex] \left(\dfrac{d^2y}{dx^2}\right)_{x=x_a} = \dfrac{1}{h^2}\left[\Delta^2 y_a - \Delta^3 y_a + \dfrac{11}{12}\Delta^4 y_a - \ldots\right]\end{array}\right\} \quad (8\text{-}24)$$

Additional terms and higher order derivatives may be readily found. Since the lower order differences disappear upon differentiation, the convergence of these expressions is not always rapid and consequently the derivatives obtained in this way are not necessarily very precise.

Maxima or minima in a tabulated function may be found by substituting the differences into the general equation given above, equating the derivative to zero and solving for u and then for x from the relationship $x = x_a + hu$.

Example. Find dy/dx and d^2y/dx^2 for $y = e^{-x^2}$ at the point $x = 0.05$ from the data given earlier in this chapter in Table 8-3.

$$\left(\frac{dy}{dx}\right)_{x=x_a=0.05} = \frac{1}{.05}\left[-0.00745 + \frac{0.00485}{2} + \frac{0.00019}{3} - \frac{0.00005}{4}\right]$$

$$= -0.09980$$

$$\left(\frac{d^2y}{dx^2}\right)_{x=x_a=0.05} = \frac{1}{(0.05)^2}\left[-0.00485 - 0.00019 + \frac{0.00055}{12}\right]$$

$$= -1.99760$$

The values obtained by ordinary differentiation are

$$\frac{dy}{dx} = -2xy = -0.099750$$

$$\frac{d^2y}{dx^2} = 2y(2x^2 - 1) = -1.985025$$

Another method for finding the derivative, first described by Rutledge, does not depend on differences but assumes that the given data can be fitted into a polynomial of the fourth or lower degree. However, five points must be known, i.e., five values of x and y. If h is the equal interval between successive values of x, the derivative of $y = f(x)$ at the point $x = x_k$ is given by the three following *approximately equivalent* expressions.

$$\left.\begin{aligned}
\left(\frac{dy}{dx}\right)_{x=x_k} &= \frac{1}{12h}[3y_{k+1} + 10y_k - 18y_{k-1} + 6y_{k-2} - y_{k-3}] \\
&= \frac{1}{12h}[(y_{k-2} - y_{k+2}) - 8(y_{k-1} - y_{k+1})] \\
&= \frac{1}{12h}[y_{k+3} - 6y_{k+2} + 18y_{k+1} - 10y_k - 3y_{k-1}]
\end{aligned}\right\} \quad (8\text{-}25)$$

These equations are particularly suitable for solution by one continuous operation with a calculating machine. The method may be extended to apply to polynomials of degree higher than four or to derivatives of higher order.

Example. Find dy/dx at $x = 0.15$ for $y = e^{-x^2}$ using the data in Table 8-3.

$$\left(\frac{dy}{dx}\right)_{x=0.15}$$

$$= \frac{1}{12 \times 0.05}[3(0.96079) + 9.7775 - 18(0.99005) + 6(0.99750) - 1]$$

$$= -0.02934$$

or

$$\left(\frac{dy}{dx}\right)_{x=0.15} = \frac{1}{0.6}[(0.99750 - 0.93941) - 8(0.99005 - 0.96079)]$$

$$= -0.02933$$

or

$$\left(\frac{dy}{dx}\right)_{x=0.15}$$

$$= \frac{1}{0.6}[0.91393 - 6(0.93941) + 18(0.96079) - 9.7775 - 3(0.99005)]$$

$$= -0.02933$$

or by direct differentiation, $dy/dx = -0.0293325$.

96. The Method of Least Squares. A situation which frequently arises in data processing that in which we wish to obtain from a number of measures of an observed quantity the best or most plausible results. The assumption is made that the differences between the observations are due to accidental errors of observation. Upon this assumption there is based a method for "reducing" the data, known as the method of least squares.

This method applies directly when the quantities observed are functions of unknowns, which are to be determined by expressing them in a series of simultaneous equations

$$\left. \begin{aligned}
a_1x_1 + b_1x_2 + c_1x_3 + \cdots n_1x_n &= k_1 \\
a_2x_1 + b_2x_2 + c_2x_3 + \cdots n_2x_n &= k_2 \\
&\cdots\cdots\cdots\cdots\cdots\cdots\cdots \\
&\cdots\cdots\cdots\cdots\cdots\cdots\cdots \\
&\cdots\cdots\cdots\cdots\cdots\cdots\cdots \\
a_px_1 + b_px_2 + c_px_3 + \cdots n_px_n &= k_p
\end{aligned} \right\} \quad (8\text{-}26)$$

where $p > n$. For if $p = n$, then the number of observations would be just enough to enable us to find the unknowns, since there would be the same number of equations as unknowns. When there are more equations than unknowns, due to the excess of observations, then the effect of accidental errors of observation means that the equations are not strictly consistent (See discussion of consistency of equations in Chapter 3), and the problem arises of satisfying them as closely as possible. This problem can be defined as that of minimizing the errors E_i, which are defined by the following equations:

$$\left. \begin{aligned}
E_1 &= a_1x_1 + b_1x_2 + c_1x_3 + \cdots n_1x_n - k_1 \\
E_2 &= a_2x_1 + b_2x_2 + c_2x_3 + \cdots n_2x_n - k_2 \\
&\cdots\cdots\cdots\cdots\cdots\cdots\cdots\cdots\cdots \\
&\cdots\cdots\cdots\cdots\cdots\cdots\cdots\cdots\cdots \\
&\cdots\cdots\cdots\cdots\cdots\cdots\cdots\cdots\cdots \\
E_p &= a_px_1 + b_px_2 + c_px_3 + \cdots n_px_n - k_p
\end{aligned} \right\} \quad (8\text{-}27)$$

The problem is simplified by omitting weighting factors, that is, by assuming that all the observations are equally reliable, an assumption that is often the most reasonable in dealing with scientific or engineering data.

Then the principle of least squares asserts that of all possible sets of values of the unknowns x_i, the most acceptable is that set which minimizes $E_i^2 = E_1^2 + E_2^2 + \cdots E_p^2$.

Therefore, we must square the set of equations (8-27) obtaining

$$
\left.
\begin{aligned}
E_1^2 &= a_1^2 x_1^2 + b_1^2 x_2^2 + c_1^2 x_3^2 + \cdots + 2a_1 b_1 x_1 x_2 + 2a_1 c_1 x_1 x_3 \\
&\quad + \cdots + 2a_1 n_1 x_1 x_n - 2a_1 k_1 x_1 - 2b_1 k_1 x_2 - \cdots + k_1^2 \\
E_2^2 &= a_2^2 x_1^2 + b_2^2 x_2^2 + c_2^2 x_3^2 + \cdots + 2a_2 b_2 x_1 x_2 + 2a_2 c_2 x_1 x_3 \\
&\quad + \cdots + 2a_2 n_2 x_1 x_n - 2a_2 k_2 x_1 - 2b_2 k_2 x_1 - \cdots + k_2^2 \\
&\quad \cdots\cdots\cdots\cdots\cdots\cdots\cdots\cdots\cdots\cdots\cdots\cdots \\
&\quad \cdots\cdots\cdots\cdots\cdots\cdots\cdots\cdots\cdots\cdots\cdots\cdots \\
&\quad \cdots\cdots\cdots\cdots\cdots\cdots\cdots\cdots\cdots\cdots\cdots\cdots \\
E_p^2 &= a_p^2 x_1^2 + b_p^2 x_2^2 + c_p^2 x_3^2 + \cdots + 2a_p b_p x_1 x_2 + 2a_p c_p x_1 x_3 \\
&\quad + \cdots + 2a_p n_p x_1 x_n - 2a_p k_p x_1 - 2b_p k_p x_2 - \cdots + k_p^2
\end{aligned}
\right\} \text{(8-28)}
$$

Then adding all equations and factoring

$$
\left.
\begin{aligned}
E_1^2 + E_2^2 \cdots E_p^2 &= (a_1^2 + a_2^2 + \cdots + a_p^2)x_1^2 + (b_1^2 + b_2^2 \\
&\quad + \cdots b_p^2)x_2^2 + (c_1^2 + c_2^2 + \cdots + c_p^2)x_3^2 + \cdots + 2(a_1 b_1 \\
&\quad + a_2 b_2 + \cdots + a_p b_p)x_1 x_2 + 2(a_1 c_1 + a_2 c_2 + \cdots + a_p c_p)x_1 x_3 \\
&\quad + \cdots + 2(a_1 n_1 + a_2 n_2 + \cdots + a_p n_p)x_1 x_n + \cdots \\
&\quad - 2(a_1 k_1 + a_2 k_2 + \cdots a_p k_p)x_1 - 2(b_1 k_1 + b_2 k_2 + \cdots \\
&\quad + b_p k_p)x_2 - \cdots + (k_1^2 + k_2^2 + \cdots + k_p^2)
\end{aligned}
\right\} \text{(8-29)}
$$

In order to minimize this sum, $\sum E_i^2$, its derivatives with respect to the unknowns, that is

$$
\frac{\partial \sum E_i^2}{\partial x_1}, \frac{\partial \sum E_i^2}{\partial x_2}, \ldots, \frac{\partial \sum E_i^2}{\partial x_n}
$$

should be equated to zero. Therefore, minimum values of the E_i are obtained by using values of the unknown x_i calculated from the equations.

$$
\left.
\begin{aligned}
(a_1^2 + a_2^2 + \cdots + a_p^2)x_1 &+ (a_1 b_1 + a_2 b_2 + \cdots + a_p b_p)x_2 \\
&+ \cdots = (a_1 k_1 + a_2 k_2 + \cdots + a_p k_p) \\
(a_1 b_1 + a_2 b_2 + \cdots + a_p b_p)x_1 &+ (b_1^2 + b_2^2 + \cdots + b_p^2)x_2 \\
&+ \cdots = (b_1 k_1 + b_2 k_2 + \cdots + b_p k_p) \\
\cdot \qquad\qquad & \qquad\qquad \cdot \\
\cdot \qquad\qquad & \qquad\qquad \cdot \\
\cdot \qquad\qquad & \qquad\qquad \cdot \\
\cdot \qquad\qquad & \qquad\qquad \cdot
\end{aligned}
\right\} \text{(8-30)}
$$

These derived equations (8-30) are called the normal equations. They are solved by the matrix methods explained in Chapter 2, or other algebraic methods, to obtain the most satisfactory values of the unknowns — which are called the *least squares adjusted values*.

Note the normal equations (8-30) may be obtained from the original equations (8-26) by multiplying each of the latter by the coefficients of the unknown in them and adding together all the products. To illustrate this method, let us solve a problem in two unknowns.

Find the least squares adjusted values of x and y from the following equations:

$$\left.\begin{array}{r} 5.49x - 13.15y = -71.24 \\ 4.84x + 8.60y = 78.56 \\ 14.36x - 12.02y = -26.48 \\ 7.42x - 3.10y = 8.62 \end{array}\right\} \quad (8\text{-}31)$$

Then to find the normal equation for x, multiply each equation in (8-31) by the coefficient of x in it, and add all the resulting equations, obtaining

$$\left.\begin{array}{r} 30.14x - 72.19y = -391.11 \\ 23.43x + 41.62y = 380.23 \\ 206.21x - 172.61y = -380.25 \\ \underline{55.07x - 23.00y = 63.96} \end{array}\right\} \quad (8\text{-}32)$$

$$314.85x - 226.18y = -327.17 \quad (8\text{-}33)$$
$$\text{(Normal equation for } x)$$

To find the normal equation for y, we use the sum of squares method as expressed in Equations (8-30). A convenient plan for applying this method systematically is first to compute all squared terms by applying (8-30) to the equations in the problem given (8-31). Thus

$$\sum a_i^2 = (a_1^2 + a_2^2 + a_3^2 + a_4^2)$$
$$= [(5.49)^2 + (4.84)^2 + (14.36)^2 + (7.42)^2]$$
$$= 30.14 + 23.43 + 206.21 + 55.06 = 314.84$$

$$\sum b_i^2 = (b_1^2 + b_2^2 + b_3^2 + b_4^2)$$
$$= [(-13.15)^2 + (8.60)^2 + (-12.02)^2 + (-3.10)^2]$$
$$= 172.92 + 73.96 + 144.48 + 9.61 = 400.97$$

$$\sum (a_i + b_i)^2 = (a_1 + b_1)^2 + (a_2 + b_2)^2 + (a_3 + b_3)^2 + (a_4 + b_4)^2$$
$$= [(5.49 - 13.15)^2 + (4.84 + 8.60)^2$$
$$+ (14.36 - 12.02)^2 + 7.42 - 3.10)^2]$$
$$= 58.68 + 180.63 + 5.48 + 18.66 = 263.45$$

$$\sum (b_i + k_i)^2 = (b_1 + k_1)^2 + (b_2 + k_2)^2 + (b_3 + k_3)^2 + (b_4 + k_4)^2$$
$$= [(-13.15 - 71.24)^2 + (8.60 + 78.56)^2$$
$$+ (-12.02 - 26.48)^2 + (-3.10 + 8.10)^2]$$
$$= 7121.67 + 7596.87 + 1482.25 + 25.00 = 16{,}225.79$$

$$\sum k_i^2 = k_1^2 + k_2^2 + k_3^2 + k_4^2$$
$$= [(-71.24)^2 + (78.56)^2 + (-26.48)^2 + (8.62)^2]$$
$$= 5075.14 + 6171.67 + 701.19 + 74.30 = 12{,}022.30$$

Now since $(a + b)^2 = a^2 + 2ab + b^2$

Therefore $ab = \frac{1}{2}[(a + b)^2 - a^2 - b^2]$

Therefore, $\sum (a_i b_i) = \frac{1}{2}[\sum (a_i + b_i)^2 - \sum a_i^2 - \sum b_i^2]$

So in this problem

$$\sum (a_i b_i) = \frac{1}{2}[263.45 - 314.84 - 400.97]$$
$$= \frac{1}{2}[-452.36] = -226.18$$

Similarly

$$\sum (b_i k_i) = \frac{1}{2}[\sum (b_i + k_i)^2 - \sum b_i^2 - \sum k_i^2]$$

So in this problem

$$\sum (b_i k_i) = \frac{1}{2}[16{,}225.79 - 400.97 - 12{,}022.30]$$
$$= \frac{1}{2}[3802.52] = 1901.26$$

Since the normal equation for y from (8-30) is

$$[\sum (a_i b_i)]x + [\sum b_i^2]y = \sum (b_i k_i)$$

So in this problem, it is

$$-226.18x + 400.97y = 1901.26 \qquad (8\text{-}34)$$
$$(\text{Normal equation for } y)$$

We can now solve the two simultaneous normal equations (8-33) and (8-34) in x and y.

$$314.85x - 226.18y = -327.17 \qquad (8\text{-}33)$$

$$-226.18x + 400.97y = 1901.26 \qquad (8\text{-}34)$$

Multiplying (8-33) by 226.18 and (8-34) by 314.85, and adding, we have

$$71{,}212.77x - 51{,}157.39y = -73{,}999.31$$

$$-71{,}212.77x + 126{,}245.40y = 598{,}611.71$$

$$75{,}099.01y = 524{,}612.40$$

$$y = 6.97$$

Substituting this value in (8-33), we have

$$x = 3.97.$$

ANSWERS TO EXERCISES AND PROBLEMS

Page 36

1. S; E; S; S.   **2.** S'; E; ϕ; E.   **3.** S; S'.

5. $\{a, b, c, d, e\}$, $\{a, b, c, d\}$, $\{a, b, c, e\}$, $\{a, c, d, e\}$, $\{a, b, d, e\}$, $\{b, c, d, e\}$, $\{a, b, c\}$, $\{a, b, d\}$, $\{a, b, e\}$, $\{a, c, d\}$, $\{a, c, e\}$, $\{a, d, e\}$, $\{b, c, d\}$, $\{b, c, e\}$, $\{b, d, e\}$, $\{c, d, e\}$, $\{a, b\}$, $\{a, c\}$, $\{a, d\}$, $\{a, e\}$, $\{b, c\}$, $\{b, d\}$, $\{b, e\}$, $\{c, d\}$, $\{c, e\}$, $\{d, e\}$, $\{a\}$, $\{b\}$, $\{c\}$, $\{d\}$, $\{e\}$, $\{\phi\}$,

6. $\{\{5\}, \{5, 7\}, \{5, 7, 12\}, \{5, 7, 12, 4\}\}$

7. dRd for every d in D **8.** dRe implies eRd

9. dRe and eRf implies dRf

10. Multiplication and (if division by 0 is excluded) division

11. $1/11$; $73112/9990000$; $142857/999999$

12. (b), (c) and (e) yield rational results

13. (a), (b), (d), and (f) are rational

14. (d), (e), (f)

17. $\begin{pmatrix} a & b & c \\ c & a & b \end{pmatrix}$ $\begin{pmatrix} a & b & c \\ a & b & c \end{pmatrix}$ $\begin{pmatrix} a & b & c \\ b & c & a \end{pmatrix}$

18. $\begin{pmatrix} a & b & c & d \\ a & d & c & b \end{pmatrix}$ $\begin{pmatrix} a & b & c & d \\ c & d & a & b \end{pmatrix}$ $\begin{pmatrix} a & b & c & d \\ d & a & c & b \end{pmatrix}$

19. There are, first, six rotation groups for rotations of 60°, 120°, 180°, 240° 300° and 360°.

$\begin{pmatrix} a & b & c & d & e & f \\ b & c & d & e & f & a \end{pmatrix}$ $\begin{pmatrix} a & b & c & d & e & f \\ c & d & e & f & a & b \end{pmatrix}$ $\begin{pmatrix} a & b & c & d & e & f \\ d & e & f & a & b & c \end{pmatrix}$

$\begin{pmatrix} a & b & c & d & e & f \\ e & f & a & b & c & d \end{pmatrix}$ $\begin{pmatrix} a & b & c & d & e & f \\ f & a & b & c & d & e \end{pmatrix}$ $\begin{pmatrix} a & b & c & d & e & f \\ a & b & c & d & e & f \end{pmatrix}$

There are also three axial reflections (inversions)

About the a-d axis   About the b-e axis   About the c-f axis

$\begin{pmatrix} a & b & c & d & e & f \\ a & f & e & d & c & b \end{pmatrix}$ $\begin{pmatrix} a & b & c & d & e & f \\ c & b & a & f & e & d \end{pmatrix}$ $\begin{pmatrix} a & b & c & d & e & f \\ e & d & c & b & a & f \end{pmatrix}$

Page 81

1. a. $\begin{pmatrix} 10 & 7 & 3 \\ 11 & 9 & 16 \end{pmatrix}$ **b.** $\begin{pmatrix} 2\frac{1}{2} \\ \frac{5}{12} \\ \frac{1}{4} \end{pmatrix}$ **c.** $\begin{pmatrix} -1 & 1 \\ 1.2 & 0 \end{pmatrix}$

2. $\begin{pmatrix} 42 & 21 & 0 \\ 7 & 35 & 28 \end{pmatrix}$

3. a. -4 b. $\begin{pmatrix} 28 & 31 \\ 48 & 51 \\ 41 & 70 \end{pmatrix}$ c. $\begin{pmatrix} -30 & -20 \\ 3-\pi & 2-2\pi \end{pmatrix}$

4. a. 19   b. $9a^3 - 120a - 218$
   c. 0; columns one and three are proportional   d. -30
5. a. $x = 7; y = 3$ b. $x = .71; y = 1.10$
   c. $x = 11; y = 3; z = 4$ d. $x = 7; y = 0; z = 3$
6. The rectangle becomes a straight line bound by the points $(0, 0)$ and $(36, 16)$. A straight line has no area, of course, and the ratio between this and that of the rectangle is indicated by the determinant of the operator being zero.
7. $\hat{A} = \begin{pmatrix} -3 & 2 & -1 \\ -7 & 8 & -5 \\ 4 & -4 & 2 \end{pmatrix}$; det $A = -2$
   Thus $A^{-1} \begin{pmatrix} 3/2 & -1 & 1/2 \\ 7/2 & -4 & 5/2 \\ -2 & 2 & -1 \end{pmatrix}$

Page 127

1. 5/12
2. $\dfrac{\binom{7}{2}\binom{2}{1} + \binom{7}{3}}{\binom{9}{3}} = \dfrac{11}{12}$
3. 1/5 (There are a total of 20 defective parts)
4. (a) 1/36   (b) 11/36 — not 1/3
5. 1/2
6. The graph will be a step function determined by the following function
$$p(X \le x) = \begin{cases} 0 \text{ if } x < 0 \\ 1/2 \text{ if } 0 \le x < 1 \\ 1 \text{ if } x \ge 1 \end{cases}$$
7. mean $= \mu_x = 480$; standard deviation $= \sigma_x = 20$
8. .972 (Binomial Formula)   9. 6/13 (Hypergeometric Formula)
10. (a) 0.143, (b) 0.053, (c) 6 or more
11. 0.423   12. 125
13. 3; 0.857

Page 159

1. a. A's strategy is 7 : 17
      B's strategy is 17 : 7
      The value of the game is $-25/24$
   b. Saddle point at -1
   c. A's strategy is 7 : 12
      B's strategy is 1 : 1
      The value of the game is 0

2. a. Saddle point at 0
   b. A's strategy is 7 : 9
      B's strategy is 0 : 3 : 0 : 13 : 0 : 0
      The value of the game is 117/16
   c. A's strategy is 7 : 6
      B's strategy is 0 : 7 : 6
      The value of the game is 3/13
3. a. Saddle point at 2
   b. A's strategy is 0 : 1 : 5
      B's strategy is 1 : 0 : 2
      The value of the game is -1
   c. A's strategy is 71 : 43 : 95
      B's strategy is 3 : 1 : 15
      The value of the game is 10/19
   d. A's strategy is 64 : 79 : 3
      B's strategy is 56 : 43 : 47
      The value of the game is 11/146

Page 200

1. Optimal solution at (12, 6) giving the linear function a value of 6. Other corner points are (4, 1), (96/5, 24/5), (32/9, 16/9).
2. The point (7, 4) is the only possible solution; the value of the linear function is 1.
3. Optimal solution at (9, 12) giving the linear function a value of -3. Other corner points at (54/7, 54/7), (13, 13), (13, 14), (12, 14), (6, 8).
4. The optimal solution is shown in the table below:

		\multicolumn{5}{c|}{Warehouses}	Factory
		W_1	W_2	W_3	W_4	Z	Excess
	F_1	60	30				90
Factories	F_2		30	60	5	20	115
	F_3				55		55
Warehouse Capacity		60	60	60	60	20	260

Min. Cost = $17.65

5. The optimal solution is:

		D_1	D_2	D_3	D_4	Factory Supply
		\multicolumn{4}{c|}{Dealers}	
	F_1	10	30	30	30	100
Factories	F_2	20				20
	F_3				15	15
Dealer Demand		30	30	30	45	

Maximum margin = $34.00.

6. Optimal solution: $x_1 = 82/17$; $x_2 = 106/17$; $x_3 = 0$; $t_1 = 0$; $t_2 = 0$.
Max. $P = 516/17$
The first tableau appears as:

	k	x_1	x_2	x_3
P	0	5	1	1
t_1	36	-1	-5	-4
t_2	40	-7	-1	-2

First pivot $= -7$
Second pivot $= -34/7$

7. Optimal solution: $x = 0$; $y = 20$; $z = 0$; $t_1 = 0$; $t_2 = 12$.　Max. $P = 80$.

Page 270

1. $\dfrac{15a}{s}$

2. $\dfrac{2}{s} + \dfrac{7}{s^2} + \dfrac{14}{s^3}$

3. $\dfrac{2}{4 + s^2} + \dfrac{s}{9 + s^2}$

4. $\dfrac{1}{2b + s}$

5. $\dfrac{c!}{(s + 2b)^{c+1}}$

6. $t \cos bt$

7. $e^{-7t/6}$

INDEX